SOIL MICROBIOLOGY

www.nipaers.com
Browse, Search, Read & Buy...
Print Books
eChapters
eBooks
Publishers
Forthcoming
Subject Catalogues
New Titles
Online Resources on:
✓ Current Affairs
✓ Reasoning, Logic & Aptitude
✓ Competitive Examinations
✓ English Language Lab
✓ Writing & Pronunciation Tools
✓ Personality Development
✓ Online Programmes for Professionals
✓ Interview Preparation
Pay Using
PhonePe
Google Pay
paytm
PayPal
UPI
CCAvenue

SOIL MICROBIOLOGY

R. Thamizh Vendan
Professor (Microbiology)
Department of Agricultural Microbiology
Agricultural College & Research Institute
Tamil Nadu Agricultural University
Madurai - 625 104, Tamil Nadu

D. Balachandar
Professor (Microbiology)
Department of Agricultural Microbiology
Tamil Nadu Agricultural University
Coimbatore - 641 003, Tamil Nadu

NEW INDIA PUBLISHING AGENCY
New Delhi – 110 034

NEW INDIA PUBLISHING AGENCY
101, Vikas Surya Plaza, CU Block, LSC Market
Pitam Pura, New Delhi – 110 034, India
Email: info@nipabooks.com
Web: www.nipabooks.com

For customer assistance, please contact
Phone: + 91-11-27 34 17 17 Fax: + 91-11- 27 34 16 16
E-Mail: feedbacks@nipabooks.com

ISBN 978-93-90083-78-7

Composed and Designed by NIPA.

TAMIL NADU AGRICULTURAL UNIVERSITY
Coimbatore-641 003, Tamil Nadu, India

Prof. N. Kumar, Ph.D.,F.H.S.I.
Vice-Chancellor

Foreword

The overarching supremacy of microorganisms in the biosphere is so evident. Despite their small size, they are certainly of immense important to man and most of them are indispensable to his survival on earth. Soil is a living system harbouring a wide range of microflora, which form a vibrant living community capable of recycling nitrogen, phosphorus and other macro and micro nutrients in the biosphere. Agriculturally important microorganisms play a vital role in augmenting and sustaining agricultural productivity. Soil inoculation with beneficial microbes such as N_2 fixers, P solubilizers and biocontrol agents proved great success.

The authors in the book have made an attempt to arrange diverse areas of soil microbiology in a concise but comprehensive manner. They have given sufficient and detailed information to inspire the students to progress to more advanced and specialized texts. I feel that a book of this type is really appropriate for under graduate and post graduate students, as it covers the University syllabus.

I congratulate Professors, Dr. R. Thamizh Vendan, and Dr. D. Balachandar of Tamil Nadu Agricultural University (TNAU) for their sincere effort in bringing out a comprehensive book on Soil Microbiology.

(N. KUMAR)
Vice - Chancellor

Coimbatore - 641 003, Tamil Nadu, India. Email: vc@tnau.ac.in

Preface

Soil Microbiology deals with fundamental aspects of microorganisms, their relationships to agriculture with an emphasis on improving the soil fertility and crop yields, biodegradation of crop residues, and combating pests and diseases. A fascinating and relatively interesting area of this Microbiology is the study of how these tiny life forms communicate and interact with each other in the soil environment. Soil microorganisms play essential roles in the nutrient cycling that are fundamental to life on the planet.

The authors have tried their best to see the syllabus of various Universities, both undergraduate and postgraduate and they feel the information presented in the book will fulfill the need. The text is clear and simple, which is explained through various diagrams. It is believed that the publication is needful not only to the students but also to the academic professionals.The authors are much thankful to their colleagues and students, who have been helpful in many ways in the preparation of the book.

Suggestions for the improvement of the book and pointing out of errors and mistakes, if any, will be greatly appreciated.

The authors are thankful to the entire team at New India Publishing Agency (NIPA), New Delhi for all the support and encouragement in bringing out this book in a nice manner.

R. Thamizh Vendan
D. Balachandar

Contents

4. **Soil air (soil gases):** A part of the soil volume which is not occupied by soil particles i.e. pore spaces are filled partly with soil water and partly with soil air. These two components (water & air) together only account for approximately half the soil's volume. Compared with atmospheric air, soil is lower in oxygen and higher in carbon dioxide, because CO_2 is continuously recycled by the microorganisms during the process of decomposition of organic matter. The soil air comes from the external atmosphere and contains nitrogen, oxygen, CO_2, and water vapour. CO_2 in soil air (0.3-1.0%) is more than atmospheric air (0.03%). Soil aeration plays an important role in plant growth, microbial population, and microbial activities in the soil.

5. **Soil microorganisms:** Soil is an excellent culture media for the growth and development of various microorganisms. Soil is not an inert static material but a medium pulsating with life. Soil is now believed to be a dynamic or living system.

Soil contains several distinct groups of microorganisms and amongst them bacteria, fungi,actinobacteria, algae, protozoa, and viruses are the most important. But bacteria are more numerous than any other kinds of microorganisms. Microorganisms form a very small fraction of the soil mass and occupy a volume of less than one percent. In the upper layer of soil (topsoil up to 10-30 cm depth i.e. Horizon A), the microbial population is very high which decreases with depth of soil. Each organism or a group of organisms are responsible for a specific change / transformation in the soil. The final effect of various activities of microorganisms in the soil is to make the soil fit for the growth and development of higher plants.

Living organisms present in the soil are grouped into two categories as follows.

1. Soil flora (microflora) eg. Bacteria, fungi, Actinobacteria, Algae and
2. Soil fauna (microfauna) animal-like eg. Protozoa, Nematodes, Earthworms, moles, ants, rodents.

Relative proportion / percentage of various soil microorganisms are: Bacteria-aerobic (70%), anaerobic (13 %), Actinobacteria(13%), Fungi /molds (03 %) and others Algae, Protozoa and viruses (0.2-0.8 %). Soil organisms play a key role in nutrient transformations.

Soil humus

Humus is the organic residue in the soil resulting from the decomposition of plant and animal residues in soil, or it is the highly complex organic residual matter in soil which is not readily degraded by the microorganisms, or it is the soft brown/dark coloured amorphous substance composed of residual organic matter along with dead microorganisms.

Composition of humus

In most soil, the percentage of humus ranges from 2-10 percent, whereas it is up to 90 percent in a peat bog. On average humus is composed of carbon (58 %), nitrogen (3-6 %, Av.5%), acids - humic acid, fulvic acid, humin, apocrenic acid, and C: N ratio 10:1 to 12:1. During their activities, the microorganisms synthesize the number of compounds that play an important role in humus formation.

Functions/Role of Humus

1. It improves the physical condition of the soil.
2. Improves the water-holding capacity of the soil.
3. Serves as a store house for essential plant nutrients.
4. Plays an important role in determining the fertility level of the soil.
5. It tends to make soils more granular with a better aggregation of soil particles.
6. Prevents leaching losses of water-soluble plant nutrients.
7. It improves the microbial/biological activity in the soil and encourages a better development of the plant-root system in the soil.
8. It acts a sa buffering agent i.e. preventa sudden change in soil pH/soil reaction.
9. Serves as a source of energy and food for the development of soil organisms.
10. It supplies both basic and acidic nutrients for the growth and development of higher plants.
11. Improves aeration and drainage by making the soil more porous.

Distinct Phases of Soil Microbiology

1. Ecological phase

Study on the quantitative and qualitative composition of the microscopic and ultramicroscopic soil population.

1

Introduction to Soil Microbiology

Soil represents a medium or substrate in which numerous microorganisms live and bring about a great variety of processes that are responsible for the continuation of the cycle of life in nature. The numerous living forms which spend their life in the soil, ranging from submicroscopic forms to the lower animal forms. With the growing recognition of the numerous processes carried out by the microorganisms in the soil there gradually emerged a branch of microbiology, which came to known as soil microbiology. It is a branch of soil science concerned with soil-inhabiting microorganisms and their functions and activities.

Since soil microbiology concerns soil microorganisms and their processes it is closely associated with soil biochemistry. Medical bacteriologists are interested in the soil as a medium for the growth and survival of disease-producing organisms. Agricultural chemists are also interested in the soil processes that result from the activities of microorganisms. General bacteriologists, zoologists, botanists are interested in a certain special group of organisms found in the soil. Recently, soil microbiology has expanded to include the study of the role of soil microorganisms in genetic engineering, in the biological control of pests and diseases, the degradation of pollutants, production and destruction of radioactive gases and its transfer. Thus microbial participation in several important processes emphasizes that soil microbiology has become a global science.

Soil Microbiology

It is a branch of science/microbiology which deals with the study of soil microorganisms and their activities in the soil.

- Deals with the microscopic organisms of the soil
- Their population and activities
- Role of microorganisms in various transformations taking place in the soil and
- Their importance in plant nutrition and crop production

Soil

It is the outer, loose material of earth's surface which is distinctly different from the underlying bedrock and the region which supports plant life. Agriculturally, the soil is the region which supports the plant life by providing mechanical support and nutrients required for growth. From the microbiologist viewpoint, the soil is one of the most dynamic sites of biological interactions in nature. It is the region where most of the physical, biological, and biochemical reactions related to the decomposition of organic weathering of parent rock take place.

Components of soil

Soil is an admixture of five major components *viz.* organic matter, mineral matter, soil-air, soil water, and soil microorganisms/living organisms. The amount/proposition of these components varieswith locality and climate.

1. **Mineral / inorganic matter:** It is derived from parent rocks/bedrocks through decomposition, disintegration, and weathering process. Different types of inorganic compounds containing various minerals are present in the soil. Amongst them the dominant minerals are silicon, aluminium, and iron, and others like carbon, calcium, potassium, manganese, sodium, sulphur, phosphorus, etc. are in trace amount. The proportion of mineral matter in a soil is slightly less than half of the total volume of the soil.

2. **Organic matter/components:** It is derived from organic residues of plants and animals added in the soil. Organic matter serves not only as a source of food for microorganisms but also supplies energy for the vital processes of metabolism which are characteristics of all living organisms. Organic matter in the soil is the potential source of N, P, and S for plant growth. Microbial decomposition of organic matter releases the unavailable nutrients into available from. The proportion of organic matter in the soil ranges from 3-6% of the total volume of soil.

3. **Soil water:** The amount of water present in the soil varies considerably. Soil water comes from rain, snow, dew, or irrigation. Soil water serves as a solvent and carrier of nutrients for plant growth. The microorganisms inhabiting the soil also require water for their metabolic activities. Soil water thus indirectly affects plant growth through its effects on soil and microorganisms. The percentage of soil-water is 25% total volume of soil.

genus *Cytophaga* and other genera (*Bacillus, Pseudomonas, Cellulomonas, Vibrio,* and *Achromobacter*) and fungal genera (*Aspergillus, Penicillium, Trichoderma, Chaetomium, Curvularia*). Lignins and proteins are partially digested by fungi, protozoa and nematodes. Proteins are degraded to individual amino acids mainly by fungi and actinobacteria. Under anaerobic conditions of waterlogged soils, methane is the main carbon-containing product which is produced by the bacterial genera (strict anaerobes) *Methanococcus, Methanobacterium,* and *Methanosarcina.*

5. Soil microbes and humus formation: Humus is the organic residue in the soil resulting from the decomposition of plant and animal residues in soil, or it is the highly complex organic residual matter in soil which is not readily degraded by microorganisms, or it is the soft brown/dark coloured amorphous substance composed of residual organic matter along with dead microorganisms.

6. Soil microbes and cycling of elements: Life on earth is dependent on cycling of elements from their elemental state to inorganic compounds, then to organic compounds and back to their elemental states. The biogeochemical process through which organic compounds are broken down to inorganic compounds or their constituent elements is known as "mineralization", or microbial conversion of complex organic compounds into simple inorganic compounds.

Soil microbes play an important role in the biochemical cycling of elements in the biosphere where the essential elements (C, P, S, N & Fe etc.) undergo chemical transformations. Through the process of mineralization organic carbon, nitrogen, phosphorus, sulphur, iron etc. are made available to the plants.

7. Soil microbes and biological N_2 fixation: The conversion of atmospheric nitrogen into ammonia by microorganisms is known as biological nitrogen fixation.

Fixation of atmospheric nitrogen is essential because of the reasons :

1. Fixed nitrogen is lost through the process of the nitrogen cycle through denitrification.
2. Demand for fixed nitrogen by the biosphere always exceeds its availability.
3. The amount of nitrogen fixed chemically and the lightning process is very less (i.e. 0.5%) as compared to biologically fixed nitrogen
4. Nitrogenous fertilizers contribute only 25% of the total world requirement while biological nitrogen fixation contributes about 60% of the earth's fixed nitrogen
5. Manufacturing of nitrogenous fertilizers by the Haber process is costly and time-consuming.

The Number of soil microorganisms carries out the process of biological nitrogen fixation at normal atmospheric pressure (1 atmosphere) and temperature (around 20 °C).

Two groups of microorganisms are involved in the process of Biological Nitrogen Fixation.

A. Non-symbiotic (free-living)

B. Symbiotic

A. Non-symbiotic (free-living)

Depending upon the presence or absence of oxygen, non-symbiotic N_2 fixing prokaryotic organisms may be classified into four types.

1. Aerobic heterotrophs (*Azotobacter, Pseudomonas, Achromobacter*)
2. Aerobic autotrophs *(Nostoc, Anabena, Calothrix,* BGA)
3. Anaerobic heterotrophs *(Clostridium, Kelbsiella, Desulfovibrio)*
4. Anaerobic autotrophs *(Chlorobium, Chromatium, Rhodospirillum Methanobacterium* etc)

B. Symbiotic (associative)

The organisms involved are *Rhizobium, Bratdyrhizobium* (legumes), *Azospirillum* (grasses), Actinobacteria *Frankia* (Casuarinas and Alder).

7. Soil microbes as biocontrol agents: Several ecofriendly bioformulations of microbial origin are used in agriculture for the effective management of plant diseases, insect pests, weeds etc. eg: *Trichoderma* sp and *Gliocladium* sp are used for biological control of seed and soil borne diseases. Fungal genera *Entomophthora, Beauveria, Metarrhizium,* and protozoa *Maltesia grandis, Malameba locustiae,* etc are used in the management of insect pests. Nuclear polyhedrosis virus (NPV) is used for the control of *Heliothis* / American bollworm. Bacteria like *Bacillus thuringiensis, Pseudomonas* are used in cotton against angular leaf spot and bollworms.

8. Degradation of pesticides in soil by microorganisms: Soil receives different toxic chemicals in various forms which cause adverse effects on beneficial soil microflora and microfauna, plants, animals and human beings. Various microbes present in the soil act as the scavengers of these harmful chemicals. The pesticide chemicals reaching the soil are acted upon by several physical, chemical, and biological forces exerted by microbes in the soil and

2. Experimental or physiological phase

Study on the physiology and biochemistry of the organisms, their role in the cycle of life in nature, and their utilization for the formation of valuable metabolic products.

3. Agronomical phase

Application of microbiological activities for the enhancement of soil fertility and crop production.

4. Pedological phase

Importance of microorganisms in soil formation and soil structure.

Scope and Importance of Soil Microbiology

Living organisms both plant and animal types constitute an important component of soil. Though these organisms form only a fraction (less than one percent) of the total soil mass, they play an important role in supporting plant communities on the earth's surface. While studying the scope and importance of soil microbiology, the soil-plant-animal ecosystem as such must be taken into account. Therefore, the scope and importance of the soil microbiology can be understood in a better way by studying aspects like

1. Soil as a living system.
2. Soil microbes and plant growth.
3. Soil microorganisms and soil structure.
4. Organic matter decomposition.
5. Humus formation.
6. Biogeochemical cycling of elements.
7. Soil microorganisms as bio-control agents.
8. Soil microbes and seed germination.
9. Biological N_2 fixation.
10. Degradation of pesticides in soil.

1. Soil as a living system: Soil inhabits a diverse groups of living organisms, both microflora (fungi, bacteria, algae and actinobacteria) and micro-fauna (protozoa, nematodes, earthworms, moles, ants). The density of living organisms in soil is very high i.e. as much as billions per gram of soil. Usually,the density of organisms is less in cultivated soil than uncultivated virgin land and the population decreases with the soil acidity. Top soil, the surface layer contains a greater number of microorganisms because it is well supplied with oxygen and nutrients. Lower layer known as subsoil is depleted with oxygen and nutrients hence it contains fewer organisms. Soil ecosystem comprises of organisms which are both, autotrophs (Algae, Blue Green Algae) and heterotrophs (fungi, bacteria). Autotrophs use inorganic carbon from CO_2 and are "primary producers" of organic matter, whereas heterotrophs use organic carbon and are decomposers/ consumers.

2. Soil microbes and plant growth: Microorganisms being minute and microscopic, they are universally present in the soil, water, and air. Besides supporting the growth of various biological systems, soil and soil microbes serve as the best medium for plant growth. Soil fauna and flora convert complex organic nutrients into simpler inorganic forms that are readily absorbed by the plant for growth. Further, they produce a variety of substances like IAA, gibberellins, antibiotics, etc. which directly or indirectly promote the plant growth.

3. Soil microbes and soil structure: Soil structure is dependent on stable aggregates of soil particles. Soil organisms play an important role in soil aggregation. Soil constituent's *viz.* organic matter, polysaccharides, lignins and gums, synthesized by soil microbes play an important role in the cementing or binding of soil particles. Further, cells and mycelial strands of fungi and actinobacteria, vermicasts from earthworm are also found to play an important role in soil aggregation.

Different soil microorganisms, having soil aggregation or soil binding properties are graded in the order as fungi >actinobacteria > gum producing bacteria > yeasts. Examples are fungi like *Rhizopus, Mucor, Chaetomium, Fusarium, Cladasporium, Rhizoctonia, Aspergillus, Trichoderma,* and bacteria like *Azotobacter, Rhizobium, Bacillus,* and *Xanthomonas.*

4. Soil microbes and organic matter decomposition: The organic matter serves not only as a source of food for microorganisms but also supplies energy for the vital processes of metabolism that are characteristics of living beings. Microorganisms such as fungi, actinobacteria, bacteria, protozoa etc. and macroorganisms such as earthworms, termites, insects, etc. play an important role in the process of decomposition of organic matter and release of plant nutrients in the soil. Thus, organic matter added to the soil is converted by oxidative decomposition to simpler nutrients for plant growth and the residue is transformed into humus. The organic matter include cellulose, lignins, and proteins (in cell wall of plants), glycogen (animal tissues), proteins, and fats (plants, animals). Cellulose is degraded by bacteria, especially those of

2

History and Development of Soil Microbiology

History of Soil Microbiology

There is enough evidence in the literature to believe that microorganisms were the earliest of the living things that existed on this planet. Man depends on crop plants for his existence and crop plants in turn depend on soil and soil microorganisms for their nutrition. Scientists from the beginning studied the microorganisms from water, air, soil, etc. and recognized the role of microorganisms in natural processes and realized the importance of soil microorganisms in the growth and development of plants.Thus, microorganisms have been playing a significant role long before they were discovered by man. Today, the soil is considered to be the main source of scavenging the organic wastes through microbial action and is also a rich storehouse for industrial microflora of great economic importance.

Unlike soil science whose origin can be traced back to Roman and Aryan times, soil microbiology emerges as a distinct branch of soil science during the first half of the 19th century. Some of the notable contributions made by several scientists in the field of soil microbiology are highlighted in the following paragraphs.

Anton Von Leeuwenhoek (1673) discovered and described microorganisms through his own made first simple microscope with a magnification of 200 to 300 times. He observed minute, moving objects which he called **"animalcules"** (small animals) which are now known as protozoa, fungi, and bacteria. He for the first time made the authentic drawings of microorganisms (protozoa, bacteria, fungi).

Robert Hook (1635-1703) developed a compound microscope with multiple lenses and described the fascinating world of the microbes.

J. B. Boussingault (1838) showed that leguminous plants can fix atmospheric nitrogen and increase the nitrogen content in the soil.

J. Von Liebig (1856) showed that nitrates were formed due to the addition of nitrogenous fertilizers in the soil.

S. N. Winogradsky discovered the autotrophic mode of life among bacteria and established the microbiological transformation of nitrogen and sulphur. Isolated for the first time nitrifying bacteria and demonstrated the role of these bacteria in nitrification (1890), further he demonstrated that free-living *Clostridium pasteurianum* could fix atmospheric nitrogen (1893). Therefore, he is considered as **"Father of Soil Microbiology"**.

W. B. Leismaan (1858) and M. S. Woronin (1866) demonstrated that root nodules in legumes were formed by a specific group of bacteria.

Jodin (1862, France) gave the first experimental evidence of elemental nitrogen fixation by microorganisms.

Robert Koch (1882) developed a gelatin plate and streak plate technique for the isolation of specific types of bacteria from the soil. He also formulated Koch's postulates to establish a causal relationship between the host - pathogen and disease.

R. Warington (1878) showed that nitrification in soil was a microbial process.

A.B.Frank (1880) discovered an actinomycetes (now remaned as actinobacteria) "*Frankia*" (Actinorhizal symbiosis) inducing root nodules in non-legumes tress of genera *Alnus sp* and *Casuarina* growing in temperate forests, coined (1885) the term "Mycorrhiza" to denote association of certain fungal symbionts with plant roots (Mycorrhiza-A symbiotic association between a fungus and roots of higher plants. Renamed the genus *Bacillus* as *Rhizobium* (1889).

H. Hellriegel and H. Wilfarth (1886) showed that the growth of non-legume plants was directly proportional to the amount of nitrogen supplied, whereas, in legumes there was no relationship between the quantity of nitrogen supplied and extent of plant growth. They also suggested that bacteria in the root nodules of legumes accumulate atmospheric nitrogen and made it available to plants. Showed that a mutually beneficial association exists between bacteria (*Rhizobia*) and legume root and legumes could utilize atmospheric nitrogen.

Beijerinck, Martinus Willem (1851-1931)

He isolated root nodule bacteria in pure culture from nodules in legumes and named them as *Bacillus radicola*. He is considered as **"The Father of Microbial ecology"**.

Beijerinck was born in Amsterdam and studied chemical engineering at the Delft Polytechnic, where, after working as a bacteriologist with an industrial company, he taught and they are degraded into non-toxic substances and thereby minimize the damage caused by the pesticides to the ecosystem. For example, bacterial genera like *Pseudomonas, Clostridium, Bacillus, Thiobacillus, Achromobacter etc. and* fungal genera like *Trichoderma, Penicillium, Aspergillus, Rhizopus, and Fusarium* are playing important role in the degradation of the toxic chemicals and pesticides in the soil.

Biodegradation of hydrocarbons: Natural hydrocarbons in soil like waxes, paraffins, oils, etc are degraded by fungi, bacteria and actinobacteria. e.g. ethane (C_2H_6), a paraffin hydrocarbon is metabolized and degraded by *Mycobacteria, Nocardia, Streptomyces, Pseudomonas, Flavobacterium* and several fungi.

Rossi (1929) and Cholondy (1930) developed the **"contact slide or buried slide"** technique for studying soil microflora.

Van Niel (1931USA) studied the chemoautotrophic bacteria and bacterial photosynthesis.

Bortels (1936) demonstrated the importance of molybdenum in accelerating nitrogen fixation by nodulating legumes.

Garrett (1936) established the school in the United Kingdom on "Soil fungi and ecological classification".

Kubo (1939, Japan) proved the role and importance of "leghaemoglobin" (red pigment) present in root nodules of legumes in nitrogen fixation.

Ruinen (1956), the Dutch microbiologist coined the term **"phyllosphere"** to denote the region of leaf influenced by microorganisms.

Alien et al (1980) suggested that VAM fungi stimulate the plant growth by physiological effects other than nutrient uptake.**Jensen (1942)** developed the method of studying nodulation on agar media in test tubes.

Barbara Mosse and J. W. Gerdemann (1944) reported the occurrence of VAM fungi (*Glomus, Acaulopora*) in the roots of agricultural crop plants which help in the mobilization of phosphate.

Starkey (1945) studied the role of bacteria (*Bacillus* and *Clostridium*) in the transformation of iron.

Barker (1945) studied the anaerobic fermentation by methane bacteria (*Methanococcus*, *Methanosarcina*)

Thornton (1947)studied root nodule bacteria from clovers.

Virtanen (1947) studied chemistry and the mechanism of leghaemoglobin in nitrogen fixation.

Nutman (1948 England) studied the hereditary mechanism of root nodulation in legumes.

Burris and Wilson (1957) developed the **"isotope technique"** to quantify the amount of nitrogen fixed and further isolated and characterized the enzyme "nitrogenase".

Bergersen (1957 Australia) elaborated on the biochemistry of nitrogen fixation in legume root nodules.

Carnham (1960 USA) discovered the nitrogen fixation by cell-free extract of *Clostridium pasteurianum*.

Date, Brockwell and Roughley (1962, Australia) developed the technique ofbio-inoculants production and seed application.

Hardy & Associates (1968, USA) developed the technique of measurement of nitrogenase activity by acetylene-reduction test coupled with gas chromatography and thereby estimation ofbiological nitrogen fixation.

R J Swaby (1970, Australia) developed "**biosuper**" containing rock phosphate sulphur and *Thiobacillus* which was used to enhance the phosphorus nutrition of plants.

Foog and Stewart (1970, UK) intensified the work on N_2 fixing blue-green algae.

Trinick (1973, Australia) isolated Rhizobia from root nodule of genus Trema (*Parasponia*) which was aunique association of *Rhizobium* with non-leguminous plants causing root nodulation.

Dobereiner and Associates (1975, Brazil) studied nitrogen-fixing potential of *Azospirillum* in some tropical forage grasses like *Digitaria*, *Panicum* and some cereals like maize, sorghum, wheat, rye etc. in their roots. He reported four species of *Azospirillum* viz. *A. lipoferum, A. brasilense, A. amazonense,* and *A. seropedica*. He coined the term **"associative symbiosis"** to denote the association between nitrogen-fixing *Azospirillum* and cereal roots. Recently this terminology has been changed and renamed as "diazotrophic biocoenocis".

Challham and Associates (1978) isolated an actinomycetous endophyte *Frankia* from root nodules of *Camptonia peregrina* which is again an example of non-leguminous root nodulation.

Dommergues & Associates (France and Senegal) discovered the stem nodules of*Sesbania rostrata* by *Azorhizobium caluinodans* which could fix nitrogen and therefore this legume can be used as an excellent green manure crop in the lowland rice cultivation. Similarly, they also discovered nitrogen-fixing stem nodules on *Casuarina* sp. caused by *Frankia*, an actinobacterium.

Louis Pasteur proved the role of soil microorganisms in biochemical changes of elements. He also showed that the decomposition of organic residues in soil was dependent on the nature of organic matter and environmental conditions.

Brefeld introduced the practice of isolating soil fungi by **"single cell"** technique and cultivating them on solid media. He used gelatin (first solidifying agent) in the culture media as a solidifying agent.

researched since 1895.He published his findings in 1898 that an agent smaller than bacteria could cause diseases, an agent that he called as a virus. In the 1880s and 1890s Beijerinck studied the disease that stunts the growth of tobacco plants and mottles their leaves (now called the tobacco mosaic virus disease). Trying to isolate a causative agent, he pressed out the juice of infected tobacco leaves and found that the juice alone was able to infect healthy plants, even after he had passed it through a filter that removed bacteria. Beijerinck was also certain that the causative agent was not a toxin because he could infect a healthy plant and from that plant infect another healthy plant, continuing this process indefinitely - therefore the infective agent had to be capable of reproduction.

He believed that the filtered juice of the infected plants was itself alive, and he called the causative agent a filterable virus (the Latin word for poison). He isolated several soil bacteria *viz.,Azotobacter chroococcum, Thiobacillus thioparus.* He isolated *Azospirillum* from *Paspalum* roots in Brazil and named it as *Azotobacter paspali* in 1922. He developed the enrichment culture technique, a way to isolate microbes with certain growth preferences. Using enrichment media, it is possible to isolate an enormous variety of bacteria with different physiological capabilities and different growth ranges. This demonstrated the impressive variety of microbial lifestyles, much broader than that of animals or plants.

Sergei Winogradsky (1856-1953)

Winogradsky extended awareness of microbial diversity. He discovered bacteria capable of **autotrophic** ("self-feeding") growth using inorganic compounds such as H_2S as their only energy source, and CO_2 as their only C-source. This was the first discovery that organisms other than green plants or algae could exist without consuming organic matter. He also discovered groups of photosynthetic bacteria that do not produce oxygen as their waste product. He developed the culture technique known as **Winogradsky column**, in which mud and water are left in glass tube exposed to light. Over time, different microbial communities grow and interchange waste products and nutrients. He established nitrogen and sulphur transformation in soil and demonstrated the role of bacteria in nitrification.

Omeliansky (1902) found the anaerobic degradation of cellulose by soil bacteria.

J. G. Lipman and P. E. Brown (1903, USA) studied the ammonification of organic nitrogenous substances by soil microorganisms and developed the **Tumbler or Beaker** for studying different types of transformation in the soil.

Hiltner (Germany, 1904) coined the term **"rhizosphere"** to denote that region of soil which is subjected to the influence of plant roots. The rhizosphere is the region where soil and plant roots make contact.

Russel and Hutchinson (1909), proved the importance of protozoa controlling/ maintaining bacterial population and their activity in soil.

Conn (1918) developed a **"direct soil examination"**technique for studying the soil microorganisms.

Rayner (1921) and Melin (1927) carried out the intensive research on Mycorrhiza.

Alexander Fleming (1881-1955)

In 1928, while working on the influenza virus, he observed that mold had developed accidentally on a *Staphylococcus* culture plate and created a bacteria-free circle around itself. He was inspired to further experiment and he found that a mold culture prevented the growth of staphylococci, even when diluted 800 times. In 1929, he discovered the antibiotic *Penicillin*, which is an important milestone in medical microbiology. Florey and Chain later isolated penicillin in pure culture. Alexander Fleming got the Nobel Prize in medicine in the year 1945. He found that natural substances are having antimicrobial activity. He reported that nasal mucous and saliva are having antimicrobial property due to the action of lysozyme. He started the **"School of soil microbiology"** at Cornell University to study microbial aspects of pesticide degradation (1961).

Selman. A.Waksman (1944)

In 1922, Waksman isolated *Thiobacillus thioxidans.* He discovered the antibiotic Streptomycin produced by *Streptomyces griseus* in the year 1944 for tuberculosis caused by*Mycobacterium tuberculosis.* Also discovered Neomycin, Actinomycin antibiotics. In 1942, he showed the importance of soil as the source of antagonistic organisms. He also studied a variety of biochemical reactions carried out by soil microorganisms while decomposing organic matter.

S. A. Waksman published the book **"Principles of Soil Microbiology"** and thereby encouraged the research in soil microbiology (1927). Studied the role of soil as the source of antagonistic organisms with special reference to soil actinomycetes (1942).

Gerretsen andMulder (Holland) studied **"phosphate mobilization"** by the soil microorganisms and showed the importance of molybdenum in nitrogen metabolism by microorganisms.

Fritch, Fogg andStewart (UK) and lyengar (India) studied nitrogen fixation by algae in general and microalgae in particular. They also intensified the work on N_2 fixing blue green algae.

James Trappe and Don Marx worked on ectomycorrhiza, colonizing the roots of forest trees.

W. S. Cook, G. C. Papavizas, J. Baker and N.S. Kerr contributed to the field of biological control of plant pathogens using antagonistic organisms from the soil. From the beginning of the 20thcentury, an emphasis was given to the study of microorganisms in soil about their physiology, ecology, interrelationship, role in soil processes, and soil fertility. A further role of fungi and actinobacteriain cellulose decomposition was better understood and cellulose decomposing, sulphur oxidizing, iron bacteria etc were isolated from soil and studied in detail.

History of Soil Microbiology in India

During the last few decades, greater emphasis has been given on some of the important aspects in soil microbiology in India which are:

1. Characterization of nitrogen-fixing *Azotobacter, Rhizobium, Beijerinckia,* blue green algae, etc.
2. Studies on phosphorus solubilizing bacteria and fungi, celluloytic microorganisms, silage production role of humic acid etc.
3. Establishment (1979) of All-India Coordinated Project (AICP) on biological nitrogen fixation (BNF) at Indian Agricultural Research Institute, New Delhi.
4. Standardization of methods of bio-inoculants application to seed and soil.
5. Seed bacterization and response of crops to bio-inoculants.

Some of the most important contributions made on the different aspects in the field of soil microbiology by the scientists and research institutes in the country are highlighted in the following paragraphs:

C. N. Acharya (1940) contributed towards the better utilization of agricultural wastes for the production of biogas and compost.

Sundara Rao (1962) established the **"Division of Microbiology"** at IARI, New Delhi.

Madhok (Punjab) introduced the practice of using bacterial cultures for berseem.

Sanyasi Rajuand Rajagopalan (Coimbatore) initiated the research work on root nodulation in legumes at Madras Agricultural College.

P. K. Dey (West Bengal) worked on free-living nitrogen-fixing organisms viz *Azotobacter, Beijerinckia* and blue green algae in rice fields and discovered nitrogen fixation by BGA in paddy.

M.O.P. lyengar (Madras Univ.) laid the foundation stone of algal research in India.

Sadasivan (Madras) and Saxena (Allahabad) studied the ecology and physiology of soil fungi along with rhizosphere phenomenon.

Singh B.N. pioneering research on soil protozoa in India.

Bhar J.V. (Bangalore) initiated work on the role of earthworms in the maintenance of soil fertility, biological nitrogen fixation and microbiology of phyllosphere.

Thirumalacher (Hindustan Antibiotics, Pune) developed the antifungal antibiotics like Haymycin and Aureofungin.

Nandi (Bose Res. Institute, Calcutta) worked on the production technology of antibiotics and bacterial fertilizers (Biofertilizers).

Desikachray (Madras)studied the taxonomy of blue green algae in India.

Thomas (BARC, Mumbai) studied the physiology of algae in India.

Raja Rammohan Rao(CRRI, Cuttak) studied on the rhizosphere nitrogen fixation phenomenon.

Bhagyaraj (GKVK, Bangalore) studied the mycorrhiza and nitrogen fixation interactions.

Verma (JNU, Delhi)studied on sulphur metabolism.

Subramaniam and Mahadevan (Univ. Madras) studied the fundamental aspects of nitrogen fixation.

Modi, Sushil Kumar, Das and Thomas carried research on the genetics of *nif* gene in relation to biological nitrogen fixationby *Rhizobium, Azospirillum* and *Kelbsiella.*

Bharadwaj (Palampur) studied on the microbiology of organic matter decomposition and therole of celluloytic microorganisms.

Gaur (IARI) and Mishra (Hissar) studied the role of celluloytic microorganisms in accelerating the process of composting and compost making.

Karla and Garcha (Ludhiana) studied the phenomenon of cellulose degradation and legume bacteriology.

Ranganathan and Neelakantan (NDRI, Karnal) worked on silage microbiology and process of anaerobic decomposition in biogas production.

Vadher, Gupta, Sethunathan and Raghu studied the role of soil enzymes and microbiology of pesticide degradation in soil.

Dart and Wani (non symbiotic nitrogen fixation), **Thomas, Kumar Rao, Nambiar and Rupela** (symbiotic nitrogen fixation) and **Krishna** (VAM fungi), these scientist at ICRISAT, Hyderabad work on nitrogen fixation in Grams, groundnut, rrhar, sorghum and millets.

N. V. Joshi (1920) reported first isolation and identification *of Rhizobium* from different cultivated legumes

Gangulee and Madhok studied physiology of *Rhizobium* and production of *Rhizobium*i noculants.

Sen and Pal (1957) studied on solubilization of phosphate by soil microorganisms.

A. Sankaran (1958) standardized the quality of legume inoculants for the first time in India.

P. K. Dey and R. Bhattacharya isolated for the first time a new, non-symbiotic nitrogen-fixing bacterium *Derixa gummosa*.

V. Iswaran (1959) reported the use of Indian peat as carrier for biofertilizers production.

Dube J. N. (1975) reported coal (woodcoal), an alternative to peat as a carrier material for the biofertilizer production.

3

Distribution and Importance of Microorganisms

Soil is a bioorganic - mineral complex of the loose surface layer of the earth's crust. Soil microbiology is the science that deals with the millions of microscopic forms of life found in the soil. It is stated that one hectare - foot (one hectare to a depth of 30cm) contains 2 to 10 metric tons of living microorganisms. Soil microorganisms are in a constant struggle for existence and are believed to be in a state of dynamic equilibrium. Without microorganisms, the soil would be an inactive geological mass incapable of supporting plant life and without plant life, animal life would not be possible. This soil population consists mainly of bacteria, fungi, actinobacteria, algae, and protozoa. The soil ecosystem includes these microbial groups as well as the inorganic and organic constituents of a given site. The collections of cells represented in the community are considered as distinct populations. All the inhabitants of the particular locality make up the community.

Types of Microorganisms in Soil

Living organisms both plants and animals, constitute an important component of soil. The pioneering investigations of several early microbiologists showed for the first time that the soil was not an inert static material but a medium pulsating with life. The soil is now believed to be a dynamic or rather a living system, containing a dynamic population of organisms/microorganisms. Cultivated soil has relatively more population of microorganisms than the fallow land, and the soils rich in organic matter contain much more population than sandy and eroded soils. Microbes in the soil are important to us in maintaining soil fertility and productivity, cycling of nutrient elements in the biosphere, and sources of industrial products such as enzymes, antibiotics, vitamins, hormones, organic acids, etc. At the same time certain soil microbes are the causal agents of human and plant diseases.

The soil organisms are broadly classified into two groups viz., soil flora and soil fauna, the detailed classification of which is as follows.

Soil Organisms

A. Soil flora

a) **Microflora:** 1. Bacteria 2. Fungi, molds, yeast, mushroom 3. Actinobacteria 4. Algae.

b) **Macroflora:** Roots of higher plants

B. Soil fauna

a) **Microfauna:** Protozoa, nematodes

b) **Macrofauna**: Earthworms. moles, ants and others.

As soil inhabit several diverse groups of microorganisms, but the most important amongst them are bacteria, actinobacteria, fungi, algae and protozoa. The characteristics and their functions in the soil are described in the next topics.

I. Bacteria

Bacteria are the most dominant group of microorganisms in soil and more numerous than the other four combined. They present in all types of soil but their population decreases as the depth of soil increases (Horizon A > B > C). The number of cells of bacteria in the soil is always great, but the individuals are small, (μm in length). Because of the minute size of the bacteria and large cells or filaments of the other 4 groups, the bacteria probably account for appreciably less than half of the total microbial cell mass. In aerated soil, bacteria will be dominating, and they are responsible for all the activities in the environment lacking O_2 or little O_2. In the transformation process, bacteria stand first, due to their rapid growth and capacity of vigorous decomposition of a variety of substrates.

Soil microbiological population has been divided into two broad groups. 1. Autochthonous 2. Zymogenous.

Autochthonous or native microbes; Indigenous, which are characteristic of the particular soil and always found to be there.eg. *Arthrobacter*

Zymogenous, or fermentative organisms require an external source of nutrients and their normal population in the soil is low (*Pseudomonas*, *Bacillus*). When specific substrates are added to the soil, the population is increased. Then gradually declines when the added substrate is exhausted. eg. Cellulose decomposers, N utilizing bacteria, nitrifiers, etc.

All the bacteria essentially need carbon, energy, and electron for their normal cellular activity, growth and reproduction. Based on the source of above, bacteria can be classified into following groups.

1. **Autotrophs:** They utilize CO_2 as a sole source for carbon requirement. Autotrophic bacteria are capable of synthesizing their food from simple inorganic nutrients, while heterotrophic bacteria depend on preformed food for nutrition. All autotrophic bacteria utilize CO_2 (from the atmosphere) as a carbon source and derive energy either from sunlight (Photoautotrophs, eg. *Chromatium. Chlorobium. Rhodopseudomonas* or from the oxidation of simple inorganic substances present in soil (Chemoautotrophs eg. *Nitrobacter, Nitrosomonas, Thiobacillus).*

2. **Heterotrophs:** They derive their carbon from organic compounds. The majority of soil bacteria are heterotrophic and derive their carbon and energy from complex organic substances or organic matter, decaying roots and plant residues. They obtain their nitrogen from nitrates and ammonia compounds (proteins) present in the soil and other nutrients from soil or the decomposing organic matter. Certain bacteria also require amino acids, vitamins, and other growth-promoting substances. Eg. Symbiotic & non-symbiotic nitrogen fixers, ammonifiers, cellulose decomposers, denitrifiers.

3. **Phototrophs:** Bacteria capable of deriving energy from sunlight.

4. **Chemotrophs:** They utilize the inorganic chemicals for their energy.

5. **Lithotroph:** The organisms utilize the inorganic substances as the electron source.

6. **Organotroph:** Bacteria capable of utilizing organic substances as the electron source.

They may exist in combinations of the above groups as follows: photolithotrophs; chemolithotrophs; photoorganotrophs; chemoorganotrophs; photoautotrophs; photoheterotrophs; chemoautotrophs; chemoheterotrophs.

Abundance of Bacteria

In Arctic regions wherein the frozen condition prevailing for 9-10 months perayear, bacterial counts of 10^6/g of soil wasobserved even when the temperature remains below freezing point. Such organisms are in a state of dormancy awaiting the spring to recover. Desert soil is another extreme. Bacteria exist even in the oven dry state. Spore forming bacilli are often predominant in this environment. Common soil bacteria are *Arthrobacter, Bacillus, Clostridium and Micrococcus*.

As per the system proposed in the Bergey's Manual of Systematic Bacteriology, most of the bacteria which are predominantly encountered in soil are taxonomically included in the three orders, Pseudomonadales, Eubacteriales and Actinomycetales of the class Schizomycetes. The most common soil bacteria belong to the genera *Pseudomonas, Arthrobacter, Clostridium, Achromobacter, Sarcina, Enterobacter* etc. Another group of bacteria common in the soilis the Myxobacteria belonging to the genera *Micrococcus, Chondrococcus, Archangium, Polyangium, Cyptophaga.*

Importance

- In transformation processes, bacteria stand first due to their rapid growth and capacity of vigorous decomposition of a variety of substrates.
- Most of the organisms are involved in N_2 fixation, P solubilization, CO_2 fixation, S and Fe transformation, Si solubilization etc.,
- They also act as the molecular vectors.

Functions of Bacteria

Bacteria bring about several changes and biochemical transformations in the soil and thereby directly or indirectly help in the nutrition of higher plants growing in the soil. The important transformations and processes in which soil bacteria play a vital role are: decomposition of cellulose and other carbohydrates, ammonification (proteins ammonia), nitrification (ammonia-nitrites-nitrates), denitrification (release of free elemental nitrogen), biological fixation of atmospheric nitrogen (symbiotic and non-symbiotic) oxidation and reduction of sulphur and iron compounds. All these processes play a significant role in plant nutrition.

Table 1: Bacterial genera involved in various processes

Process/reaction	Bacterial genera
Cellulose decomposition (Celluloytic bacteria) most cellulose decomposers are mesophilic	a. Aerobic : *Angiococcus, Cytophaga, Polyangium, Sporocytophyga, Bacillus, Achromobacter, Cellulomonas* b. Anaerobic: *Clostridium Methanosarcina, Methanococcus*
Ammonification (Ammonifiers)	*Bacillus, Pseudomonas*
Nitrification (Nitrifying bacteria)	*Nitrosomonas, Nitrobacter Nitrosococcus*
Denitrification (Denitrifies)	*Achromobacter, Pseudomonas, Bacillus, Micrococcus*
Nitrogen fixing bacteria	a. Symbiotic-Rhizobium,*Bradyrrhizobium* b. Non-symbiotic:aerobic–Azotobacter, *Beijerinckia* (acidic soils), anaerobic-*Clostridium*

Table 2 : Bacterial genera capable of degrading various plant residues in soil

Cellulose	Hemicelluloses	Lignin	Pectin	Proteins
Pseudomonas	*Bacillus*	*Pseudomonas*	*Erwinia*	*Clostridium*
Cytophaya	*Vibrio*	*Micrococcus*		*Proteus*
Spirillum	*Pseudomonas*	*Flavobacteriumm*		*Pseudomonas*
Actinomycetes	*Erwinia*	*Xanthomonas*		*Bacillus*
Cellulomonas		*Streptomyces*		

II. Fungi

As the important constituent group of the soil population, they are widely distributed in most well-cultivated soils. Fungi account for a large part of the total microbial population. Though they are not the major inhabitants, they do in fact makeup the significant part of the biomass because of the large diameter and extensive network of the filaments. Fungi exist in the soil in the form of vegetative mycelium and spores. Characteristically the fungi possess a filamentous mycelium network of individual hyphal stands. The hypha itself is rather broad and has a diameter appreciably greater than that found in the common actinobacteria. In nature, asexual spores are abundant and widespread, whereas the sexual spores are relatively uncommon.

Fungi in the soil are present as mycelial bits, rhizomorph or as different spores. Their number varies from a few thousand to a few -million per gram of soil. Soil fungi possess filamentous mycelium composed of individual hyphae. The fungal hyphae may be aseptate /coenocytic (Mastigomycotina and Zygomycotina) or septate (Ascomycotina, Basidiomycotina & Deuteromycotina).

Distribution and Abundance

As these soil fungi are aerobic and heterotrophic, they require an abundant supply of oxygen and organic matter in the soil. Fungi are dominant in acid soils because acidic environment is not suitable for the existence of either bacteria or actinobacteria. The optimum pH range for fungi lies between 4.5 to 6.5. They are also present in neutral and alkaline soils and some can even tolerate pH beyond 9.0

Several techniques have been developed for the study of the fungal flora. Plate count in burial slide technique isused for the estimation. A slight modification of pH and addition of Novobiocin and streptomycin inhibit bacteria and actinobacteria.

Estimates of fungal density revealed that a few as 20,000 to as many as 1,000,000 fungal propagules per gram. The propagule being considered as any spore, or hyphal filament that is capable of giving rise to a colony. The length of the fungal mycelium has been reported to range from 10 to 100m per g surface soil, but varies up to 500 and sometimes over 1000mt has also been obtained. Assuming the filament has an average diameter of 5μm and a specific gravity of 1.2 and taking the range of 10 to 100 m per gram. It would appear that the weight of fungi ranges from 500 to 5000 kg per ha of surface soil. Thus the filaments make up a significant part of the soil mass. They are the inhabitants of peats, flooded soils planted to rice, regions with low and extremely high salt contents, and locations in many deserts as well as sites in Antarctica. More numerous in surface layers of well-aerated and dominant in acid soils. Common genera in the soil are *Aspergillus, Mucor, Penicillium, Trichoderma, Alternaria* and *Rhizopus*.

The most commonly encountered genera of fungi in the soil are*Alternaria, Aspergillus, Cladosporium, Cephalosporium Botrytis, Chaetomium, Fusarium, Mucor, Penicillium, Verticillium, Trichoderma, Rhizopus, Gliocladium, Monilia, Pythium,* etc. Most of these fungal genera belong to the sub-division Deuteromycotina / Fungi imperfeacta which lacks sexual mode of reproduction.

Importance

- They degrade complex molecules like cellulose, hemicellulose, pectin, starch, lignin in the soil.
- Fungi improve the soil structure by forming soil aggregates.
- They cause pathogenicity in plants, animals and human beings
- Act as a predator against soil protozoa. The hyphal penetration reduces mobility and total cessation of movement.
- Nematode trapping fungi like *Arthrobotrys, Dactylaria, Dactylella, and Harposporium* trap and kill the nematodes.
- They also have a beneficial association with plant roots eg. Mycorrhizal fungi

Functions of Fungi

1. Fungi play a significant role in the soil and plant nutrition.
2. They play an important role in the decomposition of cellulose, hemi cellulose, starch, pectin, lignin in the organic matter added to the soil.

3. Lignin which is resistant to decomposition by bacteria is mainly decomposed by fungi.

4. They also serve as a food for bacteria.

5. Certain fungi belonging to sub-division Zygomycotina and Deuteromycotina are predaceous in nature and attack on protozoa and nematodes in soil and thus, maintain the biological equilibrium in the soil.

6. They also play an important role in the soil aggregation and humus formation.

7. Some soil fungi are parasitic and cause number of plant diseases such as wilts, root rots, damping-off and seedling blights eg. *Pythium, Phytophthora, Fusarium, Verticillium* etc.

8. Several soil fungi form mycorrhizal association with the roots of higher plants (symbiotic association of a fungus with the roots of a higher plant) and help in mobilization of soil phosphorus and nitrogen.

   eg. Endomycorrhiza - *Glomus*, *Gigaspora, Acaulospora,*

   Ectomycorrhiza - *Amanita, Boletus, Entoloma, Lactarius*.

III. Actinobacteria

They are unicellular like bacteria, but produce a mycelium which is non-septate (coenocytic) and more slender, like true bacteria they do not have distinct cell-wall and their cell wall is without chitin and cellulose (commonly found in the cell wall of fungi). On culture media, unlike slimy distinct colonies of true bacteria that grow quickly, actinobacteria colonies grow slowly, show powdery consistency and stick firmly to the agar surface. They produce hyphae and conidia or sporangia like fungi. Certain actinobacteria whose hyphae undergo segmentation resemble bacteria, both morphologically and physiologically.

A transitional group between the simple bacteria and fungi are the actinobacteria. These are microorganisms producing slender, branched filaments that develop into a mycelium in the soil. They are classified under bacteria, but resemble fungi in some aspects.

- Produces mycelium with extensive branching
- Like fungi, many actinobacteria form aerial mycelium and conidia
- Growth in the liquid medium does not resemble with bacteria (turbid), but form clumps and pellets like fungi.

Certain actinobacteria resemble *Mycobacterium* and *Corynebacterium* in all respects both morphologically and physiologically including susceptibility to virus attack.

Actinobacteria differ from fungi that they do not have chitin and cellulose which are commonly found in the cell walls of fungi.The colony characters are not similar to bacteria.Some species have flagella that resemble those of true bacteria. Similarities in cell wall composition and sensitivity to antibacterial compounds. Actinomycetes are sensitive to antibacterial compounds and not antifungal compounds

Common genera in soils are *Streptomyces* (70%), *Nocardia, Micromonospora, and Frankia.* They elaborate musty odour, an odour reminiscent of freshly turned soil. The metabolite of *Streptomyces* formed geosmin and other volatile products, which is responsible for the characteristic earthy odor/ smell.

Actinobacteria belonging to the order of Actinomycetales are grouped under four families *viz.*, Mycobacteriaceae, Actinomycetaceae, Streptomycetaceae and Actinoplanaceae. Actinomycetous genera which are agriculturally and industrially important are present in only two families of Actinomycetaceae and Streptomycetaceae.

Distribution and abundance

Actinobacteria are numerous and widely distributed not only in soil but also in a variety of other habitats including composts, river mud, and lake bottoms. Present in surface soil, compost, and also in lower horizons to a considerable depth. In abundance they are second only to the Bacteria and the viable counts almost equal to both. Few species can cause diseases in plants, domestic animals, and even humans. Plate count estimates give values ranging from 10^4 to 10^8 per gram of soil. They are sensitive to acidity / low pH (optimum pH range 6.5 to 8.0) and waterlogged soil conditions.

The population of actinobacteriaincreases with the depth of soil even upto horizon 'C' of soil profile. They are heterotrophic, aerobic and mesophilic (25-30°C) organisms and some species are commonly present in compost and manures are thermophilic growing at 55-65°C temperature (eg. *Thermoactinomycetes, Streptomyces*). Their population is 10^5 to 10^8perg soil in the temperate zone, but lower counts in waterlogged soils, acid peat, arctic and tundra regions. In alkaline areas, especially when dry, the relative abundance is high.

In the order of abundance in soils, the common genera of actinomycetes are *Streptomyces* (nearly 70%), *Nocardia* and *Micromonospo*ra although *Actinomycetes*, *Actinoplanes*, *Micromonospora* and *Streptosporangium* are also generally encountered.

Importance

- Involved in the decomposition of resistant components of plant and animal tissues.
- They play an active role in the transformations at high temperatures particularly in the manure and compost pits (as thermophiles are dominant).
- Cause certain soil-borne diseases of plants.
- Some actinobacteria cause infection to human and animal.
- Importance in microbial antagonism by the production of antibiotics and production of enzymes.
- Utilized in antibiotics production at a larger level.

Functions of actinobacteria

1. Decompose all sorts of organic substances like cellulose, polysaccharides, protein, fats, organic acids etc.
2. Organic residues added to the soil are first attacked by bacteria and fungi and later by actinobacteria because they are slow in activity and growth.
3. They decompose the more resistant and indecomposable organic substances and produce a number of dark black to brown pigments which contribute to the dark colour of soil humus.
4. They are also responsible for subsequent decomposition of humus (resistant material) in soil.
5. They are responsible for earthy or musty odour of freshly ploughed soils.
6. Many genera of actinobacteria synthesize a number of antibiotics like streptomycin, terramycin, aureomycin etc.
7. One of the species of actinobacteria, *Streptomyces scabies* causes "Potato scab disease".

IV. Yeast

Yeasts are non-filamentous fungi. Their presence may be demonstrated in most of the soils. Common soil yeasts are *Candida, Debaryomyces, Rhodotorula* and *Torulopsis*. Yeasts have been found in comparable numbers in soils of Antarctica, grasslands, cultivated fields, and forests.

Function and activity

- Involved in the degradation of complex molecules.
- They can utilize and degrade the major plant constituents – cellulose, hemicellulose, pectin, starch and lignin.
- Participate in humus formation from fresh organic residues.
- Carry out inorganic transformations and influences the formation of stable aggregates through hyphal penetration and the mechanical binding of the particles.

V. Algae

Algae are present in most of the soils where moisture and sunlight areavailable. Their number in soil usually ranges from 100 to 10,000 per gram of soil. They are photoautotrophic, aerobic organisms and obtain CO_2 from atmosphere and energy from sunlight and synthesize their food. They are unicellular, filamentous or colonial. Soil algae are divided into four main classes or phyla as follows:

1. Cyanophyta (Blue-green algae)
2. Chlorophyta (Grass-green algae)
3. Xanthophyta (Yellow-green algae)
4. Bacillariophyta (Diatoms or Golden-brown algae)

Out of these four classes / phyla, blue-green algae and grass-green algae are more abundant in soil. The green-grass algae and diatoms are dominant in the soils of the temperate region while blue-green algae (now renamed as Cyanobacteria) predominate in the tropical soils. Green-algae prefer acid soils while cyanobacteria are commonly found in neutral and alkaline soils. The most common genera of green algae found in soil are *Chlorella, Chlamydomonas, Chlorococcum, Protosiphon* etc. and that of diatoms are *Navicula, Pinnularia. Synedra, Frangilaria.*

Cyanobacteria are unicellular, photoautotrophic prokaryotes containing phycocyanin pigment in addition to chlorophyll. They do not possess flagella and do not reproduce sexually. They are common in neutral to alkaline soils. The dominant genera of cyanobacteria in the soil are *Chrococcus, Phormidium, Anabaena, Aphanocapra, Oscillatoria* etc. Some cyanobacteria possess specialized cells known as **heterocyst** which are the sites of nitrogen fixation. Cyanobacteria fix the nitrogen non-symbiotically in puddle paddy under water logged conditions (20-30 kg of nitrogen per ha per season). Some cyanobacteria which possess the character of symbiotic nitrogen fixation in association with other organisms like fungi, mosses, liverworts and aquatic ferns *Azolla, eg Anabaena-Azolla* association fix nitrogen symbiotically in rice fields.

They are photosynthetic organisms and abundant in habitats, where adequate moisture and accessible light are available. Isolation can be done in liquid media and by plating with agar media. Because of their similarities and evolutionary relation to bacteria, cyanobacteria are classified together with eubacteria.

Photoautotrophic nutrition

In the absence of light heterotrophy occur in some species of Chlorophyta, Cyanophyta and Diatoms. These species use the oxidation of organic carbon to replace the light in supplying energy for the anabolic process. (Facultative photoautotrophs).

The occurrence of this photosynthetic microorganisms has been recorded in soils throughout the world. Algae are moderately adaptable to environmental changes; persist in diverse ecosystems such as in alkaline and desert soils. Flooded paddy field provides a conducive environment for the growth of cyanobacteria and green algae. The common inhabitants are *Anabaena, Nostoc, Calothrix* and *Oscillatoria*.

Importance

- Involved in CO_2 fixation.
- Colonizes the barren surface and corrode and weather rocks which contribute to soil formation.
- The algal layer covers the rocks, on death it favors secondary colonizers.
- Surface algal bloom reduces erosion losses probably by binding together with soil particles.
- Improve soil structure, texture and add fertility to the soil after decay.
- Photosynthesis liberates $O_{2,}$ provides O_2 to the roots and other microbes.
- N_2 fixation and ammonia excretion.
- N_2 is released after decomposing the algal biomass.
- Production of metabolites.
- Single-cell protein and β carotene production

Functions of algae and cyanobacteria

1. Plays an important role in the maintenance of soil fertility especially in tropical soils.
2. Add organic matter to the soil when die and thus increase the amount of organic carbon in soil.
3. Most of the soil algae (especially cyanobacteria) act as cementing agent in binding soil particles and thereby reduce the soil erosion.
4. Mucilage secreted by the algae and cyanobacteria is hygroscopic in nature and thus helps in increasing the water retention of soil for a longer period.
5. Soil algae through the process of photosynthesis liberate a large quantity of oxygen in the soil environment and thus facilitate the aeration in submerged soils.
6. They help in checking the loss of nitrates through leaching and drainage especially in un-cropped soils.
7. They help in the weathering of rocks and building up the soil structure.

VI. Protozoa

The subterranean earth contains protozoa, earthworms, nematodes, and insects. The most abundant one is protozoa, the simple form of animal life. Most of the soil forms are flagellates, amoebae or ciliates. Primitive, unicellular and majority of them are dependent on preformed organic matter for their nutrition from soluble organic and inorganic substances, or by phagotrophic nutrition (direct feeding upon microbial cells). They are regulating the biological equilibrium in soil by engulfing the bacteria. They also participate in the decomposition of plant remains.

These are unicellular, eukaryotic, colourless, and animal like organisms (Animal kingdom). They are larger than bacteria and size varying from few microns to a few centimeters. Their population in arable soil ranges from 10,000 to 1,00,000 per gram of soil and is abundant in surface soil. They can withstand the adverse soil conditions as they are characterized by the cyst stage in their life cycle. Except for few genera which reproduces sexually by fusion of cells, the rest of them reproduce asexually by binary fission. Most of the soil protozoa are motile by flagella or cilia or pseudopodia as locomotory organs.

Depending upon the type of appendages provided for locomotion, protozoa are

1. Rhizopoda (Sarcodina)
2. Mastigophora
3. Ciliophora (Ciliata)
4. Sporophora (not common inhabitants of soil)

Class-Rhizopoda: It consists of protozoa without appendages usually have naked protoplasm without cell-wall. Some times they have pseudopodia as temporary locomotory organs. Important genera are *Amoeba, Biomyxa, Euglypha,* etc.

Class-Mastigophora: It belongs to flagellated protozoa, which are predominant in soil. Important genera are *Allention, Bodo, Cercobodo, Cercomonas, Entosiphon Spiromonas, Spongomions* and *Testramitus.* Many members are saprophytic and some possess chlorophyll and are autotrophic. In this respect, they resemble unicellular algae and hence are known as "Phytoflagellates".

Class-Ciliophora: The soil protozoa belonging to the class ciliate / ciliophora are characterized by the presence of cilia (short hair-like appendages) around their body, which helps in locomotion. The important soil inhabitants of this class are *Colpidium, Colpoda, Balantiophorus, Gastrostyla, Halteria, Uroleptus, Vortiicella, Pleurotricha* etc.

Protozoa are abundant in the upper layer (15 cm) of soil. Organic manures, soil moisture, aeration, temperature and pH are the important factors affecting soil protozoa.

Functions of protozoa

1. Most protozoans derive their nutrition by feeding or ingesting soil bacteria such as *Enterobacter, Agrobacterium, Bacillus, Escherichia, Micrococcus,* and *Pseudomonas* and thus, they play an important role in maintaining the microbial equilibrium in the soil.

2. Some protozoa have been recently used as biological control agents against phytopathogens.

3. Species of the bacterial genera viz. *Enterobacter* and *Aerobacter* are commonly used as the food base for isolation and enumeration of soil protozoans.

4. Several soil protozoa cause diseases in human beings which are carried through water and other vectors, eg. Amoebic dysentery caused by *Entamoeba histolytica.*

Importance of soil microorganisms

Soil microorganisms are very important as almost every chemical transformation taking place in the soil involves active contributions from these organisms. In particular, they play an active role in soil fertility as a result of their involvement in the cycle of nutrients like carbon and nitrogen, which are required for plant growth. For example, soil microorganisms are responsible for the decomposition of the organic matter entering the soil and therefore in the recycling of nutrients. Certain soil microorganisms such as mycorrhizal fungi can also increase the availability of mineral nutrients (e.g. phosphorus) to plants. Other soil microorganisms can increase the amount of nutrients present in the soil. For instance, nitrogen-fixing bacteria can transform nitrogen gas into soluble nitrogenous compounds that plant roots can utilize for growth. These microorganisms, which improve the fertility status of the soil and contribute to plant growth, have been termed as **biofertilizers** and are receiving increased attention for use in agriculture. Similarly, other soil microorganisms have been found to produce compounds (such as vitamins and plant hormones) that can improve plant health and contribute to higher crop yield. These microorganisms (called phytostimulators) are also studied for possible use as microbial inoculants to improve crop yield.

In contrast to these beneficial soil microorganisms, other soil microorganisms are pathogenic to plants and may cause considerable damage to crops. Large numbers of pathogenic microorganisms are routinely found in the soil and many of them can infect the plant through the roots. However, certain native microorganisms present in the soil are antagonistic to these pathogens and can prevent the infection of crop plants. Antagonism against plant pathogens usually involves competition for nutrients and/or production of inhibitory compounds such as secondary metabolites (antimicrobial metabolites and antibiotics) and extracellular enzymes. Other soil microorganisms produce compounds that stimulate the natural defense mechanisms of the plant and improve its resistance to pathogens. Collectively, these soil microorganisms have been termed as biopesticides and represent an emerging and important alternative (i.e. biological control) to the use of chemical pesticides for the protection of crops against certain pathogens and pests.

4

Factors Affecting Microbial Activities in Soil

Several environmental conditions affect the density and composition of the microflora and frequently alter their activities in soil. Primary factors include soil moisture, aeration, temperature, pH, organic matter and inorganic fertilizers. Lesser variables are the secondary factors, which include crop rotation, season, soil depth, cultivation practices etc.

I. Primary Factors

Soil moisture

Soil moisture is one of the important factors influencing the microbial population. Water is the major component of protoplasm; an adequate supply must be available for vegetative development of organisms. But, when it becomes excessive, proliferation is suppressed due to limitation in gaseous exchange and lowers the availability of O_2 supply creating an anaerobic environment. Most of the organisms prefer a moisture percentage between 20 and 60 per cent. Many bacteria and fungi can adjust themselves to different moisture conditions. Under dry conditions, the bacteria may form spores that can resist the drought conditions. The fungi may sporulate or form chlamydospores to tide over adverse conditions. The protozoans also may form cysts and can survive under dry conditions. Actinobacteria are the chief group of organisms that prefer dry conditions. It is believed that the concentration of nutrients is diluted and also the aeration is very much limited at moisture and hence only the anaerobic and microaerophilic organisms can develop better.

Soil air

It is directly linked up the moisture level of the soil. Most of the forms are active in aerated soil and under waterlogged soils, anaerobic and microaerophilic forms develop.

Soil temperature

It exhibits considerable influence on the microbial population. Though microorganisms have been found to exist under extreme temperature conditions, such as – 60° C and + 60° C the soil temperature usually does not reach such extremes. The microbial population varies both quantitatively and qualitatively under extreme conditions. In tropical and subtropical regions, temperatures vary widely in summer and winter seasons and the population may also be varied. In temperate regions, there is not much variation in temperatures of seasons and not much variation in the soil population. Soil temperature influences the temperature of the air, water and solid phase of the soil. Soil water and temperature exert a combined influence on the microbial population.

Soil organic matter

Community size is related to the organic matter content so that humus-rich localities have the largest biological numbers. The organic matter content varied with soil types from less than 0.5 percent in desert soils to 40 percent in peaty soils. As the organic matter is the chief source of energy and food for most soil organisms, it has a great influence on the population. The nature of organic matter is responsible for the differential stimulation of the population. There are several indirect effects of the organic matter on soil microflora. It influences the structure and texture of the soil, besides enriching with nutrients for plants and microorganisms. Such influences on the soil also greatly affect the activity of the soil microorganisms.

Soil pH

It is a key factor influencing the microflora of soil. It influences the enzyme system and thus plays an important role in microbial activity. In general, fungi thrive better than bacteria and actinobacteria in acid soils. Bacteria flourish well in neutral and alkaline soils. The saline and alkaline soils have different microflora. Salinity is due to excess salts; alkalinity is due to high H-ion concentration. Several direct and indirect effects of H-ion and salt concentration in the soil are exerted on the microbial population.

Fertilizer application

The application of fertilizers to the soil improves the microbial activity because of the availability of more readily obtainable nutrients. Some fertilizers may however have an inhibitory effect on specific bacterial types. The continuous application of ammoniacal fertilizers favours the growth of fungi due to the formation of nitric acid that inhibits the growth of bacteria and actinobacteria.

The addition of nitrate inhibits the activity of free-living N_2 fixing bacteria like *Azotobacter*. Some of the autotrophs are encouraged by the addition of fertilizers.

Cropping and vegetation

Crop plants exhibit two kinds of effects. One is through root exudates, which may have different compounds with reference to the crops grown. Vegetation has selective stimulation over the microbial population. Continuous cultivation leads to more microbial activity than the uncultivated land.

II. Secondary Factors

Crop rotation

Crop rotation with different species like legumes, graminaceous plants, etc. brings about different stimulatory effect on the microflora. Some crops have deeper roots than others and some are more fibrous. Such variations bring about physical changes in soil, which in turn may have direct and indirect stimulatory effects on the soil microflora.

Cultural practices

Various cultural practices, such as tillage operations and irrigation have several physical and chemical changes in soil, which are reflected in the soil microflora. Through subsoil ploughing deeper layers may get better aeration and there may be quicker multiplication of aerobic organisms. Weeding, irrigation etc. may influence the microbial populations in the soil.

Soil depth

Most of the organisms are almost in top layers, largely in upper few centimeters and decline with greater depth, more active at a few centimeters down and less active at deeper layers. Low O_2 and less sunlight in deeper layers reduce the soil microbial population.

Season

Cell numbers are greatest during the spring and autumn and a decline occurs during hot, dry and winter, the cells remain in a state of dormancy for biochemical inactivity. The influence is mainly due to the alterations in temperature, moisture as well as the availability of organic matter.

III. Specific influences

- Sunlight favours algae and other autotrophs.
- Herbicide application has a devastating effect on algae.
- Attack by neighbours – protozoa, nematode, earthworms consume algae.
- Presence of parasites and predators eg. viruses eat on specific bacteria
- An increase in the bacterial population increases protozoans also.

IV. Other Factors

Burning of topsoil

- This leads to partial sterilization of the topsoil and may kill the protozoan population, which may lead to an increase in the bacterial population. This condition may affect biological equilibrium in soil.
- Application of nematicide, fungicide, and bactericide may exhibit partial sterilization.Thus, every change in crop production directly or indirectly alters the soil microflora, and still, many more are not clearly understood.

5

Factors Affecting Distribution and Population of Soil Microbes

Soil microorganisms (Flora & Fauna) are just like higher plants depends entirely on soil for their nutrition, growth and activity. The major soil factors which influence the microbial population, distribution and their activity in the soil are

1. Cultural Practices

1. Soil fertility
2. Soil moisture
3. Soil temperature
4. Soil aeration
5. Light
6. Soil pH (H-ion Concentration)
7. Organic matter
8. Food and energy supply
9. Nature of soil
10. Microbial associations

All these factors play a great role in determining not only the number and type of organism but also their activities. Variations in any one or more of these factors may lead to the changes in the activity of the organisms which ultimately affect the soil fertility level. A brief account of all these factors influencing soil micro flora and their distribution are discussed here.

1. Cultural practices (Tillage): Cultural practices viz., cultivation, crop rotation, application of manures and fertilizers, liming and gypsum application, pesticide/ fungicide and weedicide application have their effect on soil organisms. Ploughing and tillage operations facilitate aeration in soil and exposure of soil to sunshine and thereby increase the biological activity of organisms, particularly bacteria.

Crop rotation with legume maintains the favourable microbial population balance, particularly of N_2 fixing bacteria and thereby improve soil fertility.

- Liming of acid soils increases activity of bacteria and actinobacteria and lowers the fungal population. Fertilizers and manures applied to the soil for increased the crop production, supply food and nutrition not only to the crops but also to the microorganisms present in the soil and thereby proliferate the activity of microbes.
- Foliar or soil application of different chemicals (pesticides, fungicides, nematicides etc.) in agriculture are either degraded by the soil organisms or are liable to leave the toxic residues in the soil which are hazardous to cause a profound reduction in the normal microbial activity in the soil.

2. Soil fertility: The fertility level of the soil has a great influence on the microbial population in soil. The availability of N, P and K required for plants as well as microbes in the soil determines the fertility level of the soil. On the other hand, soil microflora has a greater influence on the soil fertility level.

3. Soil moisture: It is one of the important factors influencing the microbial population & their distribution in the soil. Water (soil moisture) is useful to the microorganisms in two ways i.e. it serves as a source of nutrients and supplies hydrogen oroxygen to the organisms and it serves as a solvent and carrier of other food nutrients to the microorganisms. Microbial activity andpopulation proliferate best in the moisture range of 20% to 60%. Under excess moisture conditions or water logged conditions due to lack of soil aeration (oxygen) anaerobic microflora becomes active and the aerobes get suppressed. While in the absence of adequate moisture in the soil, some microbes die out due to tissue dehydration and some of them change their forms into resting spores or cysts and tide over adverse conditions. Therefore optimum soil moisture (range 20 to 60 %) must be there for a better population of microbes in the soil.

4. Soil temperature: Next to moisture, the temperature is the most important environmental factor influencing the biological, physical and chemical processes of microbes, microbial activity and population in the soil. Though microorganisms can tolerate extreme temperature (such as - 60 °C or + 60° C) conditions, the optimum temperature range at which soil microorganisms can grow and function actively is rather narrow.

- Depending upon the temperature range at which microorganisms can grow and function are divided into three groups i.e. Psychrophiles (growing at low temperature below 10 °C), Mesophiles (growing well in the temp range of 20 ° C to 45° C) and Thermophiles (can tolerate temperature above 45° C and optimum 45-60°C).

- Most of the soil microorganisms are mesophilic (25 to 40 °) and optimum temperature for most mesophiles is 37° C. True psychrophiles are almost absent in soil, and thermophiles though present in soil behaves like mesophiles. True thermophiles are more abundant in decaying manure and compost heaps where high temperature prevails.
- Seasonal changes in soil temperature affect the microbial population and their activity especially in the temperate regions. In winter, when the temperature is low, the number and activity of microorganisms fall, and as the soils warm up in spring, they increase in number as well as activity. In general, the population and activities of soil microorganisms are the highest in spring and lowest in the winter season.

5. Soil air (Aeration): For the growth of microorganisms better aeration (oxygen and sometimes CO_2) in the soil is essential. Microbes consume oxygen from soil air and give outcarbon dioxide. Activities of soil microbes are often measured in terms of the amount of oxygen absorbed or the amount of CO_2 evolved by the organisms in the soil environment. Under high soil moisture level or waterlogged conditions, the gaseous exchange is hindered and the accumulation of CH_4occurs in soil air which is toxic to microbes. Depending upon oxygen requirements, soil microorganisms are grouped into categories viz., aerobic (require oxygen for life processes), anaerobic (do not require oxygen), and microaerophilic (requiring low concentration of oxygen).

6. Light: Direct sunlight is highly injurious to most of the microorganisms except algae. Therefore the upper portion of the surface soil a centimeter or less is usually sterile or devoid of microorganisms.

7. Soil reaction / soil pH: Soil reaction has a definite influence on quantitative and qualitative composite on the soil microbes. Most of the soil bacteria, cyanobacteria, diatoms and protozoa prefer a neutral or slightly alkaline reaction between pH 4.5 and 8.0 and fungi grow in an acidic reaction between pH 4.5 and 6.5 while actinobacteria prefer slightly alkaline soil reactions. Soil reactions also influence the type of bacteria present in the soil. For example nitrifying bacteria (*Nitrosomonas & Nitrobacter*) and diazotrophs like *Azotobacter* are absent totally or inactive in acid soils, while diazotrophs like *Beijerinckia, Derxia,* and sulphur oxidizing bacteria like *Thiobacillus thiooxidans* are active in acidic soils.

8. Soil organic matter: The organic matter in soil being the chief source of energy and food for most of the soil organisms, it has a great influence on the microbial population. Organic matter influences directly or indirectly on the population and activity of soil microorganisms. It influences the structure and texture of soil and thereby the activity of the microorganisms.

9. Food and energy supply: Almost all microorganisms obtain their food and energy from the plant residues or organic matter added to the soil. Energy is required for the metabolic activities of microorganisms. The heterotrophs utilize the energy liberated during the oxidation of complex organic compounds in soil, while autotrophs meet their energy requirement from the oxidation of simple inorganic compounds (chemoautotroph) or solar radiation (Photoautotroph). Thus, the source of food and energy-rich material is essential for the microbial activity in the soil. The organic matter, therefore, serves both as a source of food nutrients as well as the energy required by the soil organisms.

10. Nature of soil: The physical, chemical and physicochemical nature of the soil and its nutrient status influence the microbial population both quantitatively and qualitatively. The chemical nature of the soil has a considerable effect on the microbial population in the soil. The soils with the good physical condition have better aeration and moisture which is essential for optimum microbial activity. Similarly macro- and micronutrients and organic constituents of humus are responsible for the absence or presence of a certain type of microorganisms and their activity. For example activity and presence of nitrogen-fixing bacteria are greatly influenced by the availability of molybdenum and the absence of available phosphate restricts the growth of *Azotobacter.*

11. Microbial interactions: Microorganisms interact with each other giving rise to antagonistic or symbiotic interactions. The association existing between one organism and another whether of symbiotic or antagonistic influences the population and activity of soil microbes to a great extent. The predatory habit of protozoa and some mycobacteria which feed on bacteria may suppress or eliminate certain bacteria. On the other hand, the activities of some of the microorganisms are beneficial to each other. For instance organic acids liberated by fungi, increase in oxygen by the activity of algae, change in soil reaction, etc. favors the activity of bacteria and other organisms in the soil.

12. Root exudates: In the soil where plants are growing, the root exudates also affect the distribution, density, and activity of the soil microorganism. Root exudates and sloughed off materials of root surface provide an abundant source of energy and nutrients and thus directly or indirectly influence the quality as well as quantity of microorganisms in the rhizosphere region. Root exudates contain sugars, organic acids, amino acids, sterols, vitamins, and other growth factors that have a profound effect on soil microbes.

Methods for elucidating microbial biomass of soil

More recently, newer methods have been developed that use measures of components of all cells to estimate biomass. These methods include:

a) Dilution plate counts and direct microscopical counting

Dilution plates are usually able to culture 10% of the viable organisms in soil samples. Direct microscopic observation methods (FITC, acridine orange staining, etc.) usually overestimate the number of cells because they include dead organisms or other particles in their count.

b) Determination of cell wall components of bacteria

Bacteria contain specific cell wall components such as muramic acid that can be released by acid hydrolysis and analyzed by High-Pressure Liquid Chromatography (HPLC). The amount of muramic acid in bacterial cell walls varies depending on whether the cells are Gram-positive or Gram-negative.

c) Soil fumigation method

The fumigation of soil with $CHCl_3$ followed by extraction with various salt solutions has been used to measure microbial biomass.

d) Determination of ATP content of soils

Adenosine triphosphate (ATP) is an important energy compound in the metabolism of all living organisms. ATP is considered to be a useful indicator of life in the soil. ATP can be measured with great sensitivity. Though dependent on the physiological state, ATP concentrations are fairly uniform relative to cell carbon for many bacteria, algae, and protozoa. ATP is an accepted indicator of microbial biomass in aquatic and terrestrial environments. Fireflies or lightning bugs make light within their bodies. This process is bioluminescence and is shared by many other organisms, mostly sea living organisms. The cells contain a chemical called luciferin and make an enzyme called luciferase. To make light, the luciferin combines with oxygen to form an inactive molecule called oxyluciferin. The wavelength of light given off is between 510 and 670 nm (pale reddish-green color). The amount of light emitted in this reaction is directly proportional to the ATP concentration. Measurements of ATP can be used to estimate microbial biomass.

6

Microbial Biomass in Soil Fertility

Soil microbial biomass is an important parameter linking the plants to soil. It comprises about 2-3% of the total organic carbon in the soil and recognized as an important source of nutrients to plants. It is also a dynamic component of the soil associated with several functional properties of the terrestrial ecosystem. It represents the fraction of the soil responsible for the energy and nutrient cycling and the regulation of organic matter transformation.

Microbial biomass is a small but labile and living component of soil organic matter that is involved in most bio-geo-chemical processes in the terrestrial ecosystem. It interacts with crop productivity by regulating nutrient availability as a source; determining soil carbon storage- sink; and contributing to the atmospheric CO_2 from respiration. Thus it is the main agent that controls the flow of carbon and cycling of nutrient elements.

There is a close relationship between soil microbial biomass, decomposition rate, and nitrogen mineralization. It is also positively correlated with grain yield in organic farming and also contributes to soil structure and stabilization. Thus, there is the immense importance of microbial biomass in soil fertility as well as in agriculture.

Definition

Soil microbial biomass is the living component of soil organic matter excluding soil animals and plant roots. In other words it can be defined as the concentration of microbes in the soil. It is an important indicator of soil health.

Composition of Microbial Biomass

Soil microbial biomass consists of bacteria, fungi, actinobacteria, protozoa, etc. However, bacteria and fungi are the dominant organisms with regards to biomass and metabolic activities. According to Sylvia *et al.* (1998), the numbers of microbes in the soil are as follows:

Microbial group	No. per gram of soil
Bacteria	100,000,000 -1,000,000,000
Fungi	100,000 - 1,000,000
Actinobacteria	1000 - 1,000,000
Protozoa	1000 - 100,00

Role of Soil Microbial Biomass

Although it comprises less than 5% of organic matter in the soil, it performs at least 3 critical functions for plant production in the ecosystem.

- It is a labile source of carbon (C), nitrogen (N), phosphorus (P), and sulfur (S)
- It is an immediate sink of C, N, P, and S
- It is an agent of nutrient transformation and pesticide degradation.

Also, microorganisms form symbiotic associations with roots, act as biological agents against plant pathogens, contribute towards soil aggregation, and participate in soil formation.

Importance of microbial biomass in soil fertility

The importance of microorganisms in ecosystem functioning has led to an increased interest in soil microbial biomass. The importance of microbial biomass in soil fertility can be described in the following:

- As a source of nutrients
- As a sink of nutrients
- As an indicator of heavy metals and pesticides pollution
- As an indicator of soil quality.

As a source of nutrients

- It acts as a source of potentially mineralisable nitrogen
- It also acts as a source of available phosphorus and sulfur

As a sink of nutrients

Grasslands, woodlands, and forests contain substantial amounts of microbial biomass. For example, the average microbial biomass pool size is 20 - 50% higher in grasslands and forest soils than in arable soils, mainly because of the larger organic C inputs in the former. The addition of easily metabolizable C to soil rapidly increases microbial biomass, often by several folds, and also incorporates nutrients into microbial biomass rapidly, often within a week of addition.

However, such a microbial pool size is transient, and the level of microbial biomass a soil can sustain under a steady-state condition depends on the protective capacity of the soil, which is governed by a dynamic interaction between clay and clay aggregates, soil microbial biomass and soil organic matter and possibly climate.The soil microbial biomass has an important role in soil fertility can be established by the evidence of many workers. They indirectly measured the amount of nutrient elements in the microbial biomass of twenty six soils. The average amount of N and P stored in the vegetative cells of the microflora of arable soil was estimated to be about 108 and 83 kg/ha respectively.

As an indicator of heavy metals and pesticides pollution

Increasing the disposal of sewage sludge on land has the potential to increase heavy metals in soil. There is thus concern about the long-term effects of additions of heavy metals in sewage sludge to the soil. It is believed that microbial biomass, being the living component of soil, can provide an early indication of the adverse effects of increasing concentrations of heavy metals and pesticides in soil. The microbial biomass decreases faster as the heavy metal concentration increases. The microbial quotient decreases much faster than the total microbial biomass as the heavy metal concentration increases.

As an indicator of soil quality

Soil quality is the capacity of a specific kind of soil to function, within natural or managed ecosystem boundaries, to sustain plant and animal productivity, maintain or enhance water and air quality, and support human health and habitation. To evaluate soil quality, the capacity of the soil to function needs to be measured using appropriate indicators.

Microbial biomass performs 3 essential functions in soil. It is also sensitive enough to measure the changes due to the different land use and management such as virgin land brought under cultivation, crop rotations, restoration, and heavy metal contamination. However, currently, microbial biomass does not provide a benchmark, critical, or threshold values against which soil quality can be evaluated. To overcome the problem, some microbial indices for soils of different microbial biomass contents have been developed, such as:

a) **Microbial quotient:** it is the ratio of biomass C to total organic carbon.

b) **Metabolic quotient:** microbial respiration rate per unit of microbial biomass. It is also known as a specific respiratory rate.

Factors Affecting the Amount of Soil Microbial Biomass

1. Temperature and moisture
2. Soil matrix
3. Land use
4. Crop rotations
5. Fertilizers and soil amendments
6. Tillage practices

Methods to Estimate Soil Microbial Biomass

In the past, interest in microbial biomass assessment in soil has been limited by tedious, time consuming and unreliable techniques (direct microscopy, culture media) for enumeration of individual microbial communities.Estimation of microbial biomass from actual enumeration of the whole suite of microorganisms (e.g. bacteria, fungi, actinobacteria, protozoa, nematodes) has been difficult because of incomplete extraction, inappropriate growth media, extremely variable growth habits of microorganisms in the soil, and scant information on their biovolumes. In the last 20 years, relatively rapid assessment of soil microbial biomass has been possible based on physiological, biochemical, and chemical techniques. These methods can be broadly divided into direct and indirect methods.

Direct methods

a) Phospholipid fatty acids (PLFAs) analysis

b) Adenosine triphosphate (ATP) analysis

Indirect methods

a) Chloroform fumigation incubation (CFI) method

b) Chloroform fumigation extraction (CFE) method

c) Substrate-induced respiration (SIR) method

Out of these methods, Chloroform fumigation incubation (CFI) method and Chloroform fumigation extraction (CFE) method have been widely used to estimate microbial biomass in agricultural, pastoral and forestry systems, rehabilitation of disturbed lands, and pesticide and heavy metals polluted lands and materials.Among the methods, CFE method is most popular and used extensively, because of its cheaper, faster, and easier method for estimating soil microbial biomass.

Estimation of soil microbial biomass

Soil serves as the rich medium for the growth of microorganisms. The soil organic matter content has a direct bearing over the soil microbial population as they mineralize the organic matter and release the nutrients from complex organic polymers. Hence quantification of soil microbial population or microbial biomass in the soil will reflect on the fertility of soil.

In the 'Soil Fumigation Incubation Technique' method the soil is fumigated with a volatile fumigant (chloroform) followed by its removal and inoculation with fresh soil. The fumigated soil leads to an increase in the CO_2 evolution/O_2 consumption due to the mineralization of dead microbial cells over an unfumigated soil. A comparison of these values will lead to the quantification of soil microbial biomass. The flush of mineralization is caused by the flush in the decomposition of organisms killed during fumigation. Therefore the size of the flush is related to the size of the soil microbial biomass.

7

Soil Enzymes and Their Roles

Soil fertility depends on the physical, chemical, and biological properties of the soil. Among the biological properties, soil enzymes are used as an index of soil fertility. Soil enzymes respond to changes in soil management faster in comparison to other soil variables, and also catalyze the reactions necessary for the life processes of microorganisms in soils. The biochemical reactions are brought about by the catalytic action of enzymes. Soil enzymes include amylase, phosphatase, urease, cellulase, protease, glycosidase, and arylsulfatase. They play a role in organic material decomposition, the transformation of organic matter, nutrient cycling, nitrogen fixation, detoxification of pesticides, xenobiotics, etc., and thus regulate the ecosystem.

Depending on their location, enzymes can be extracellular or intracellular. Intracellular enzymes are found in the cell's cytoplasm or bound to the cell walls of living and metabolically active cells. Extracellular enzymes released into the soil and are "permanently" immobilized on clay and humic colloids via ionic interactions, covalent bonds, hydrogen bonding, entrapment, and other mechanisms. There are two kinds of soil enzymes.

1. Constitutive Enzymes

It is always present in nearly constant amounts in a cell (not affected by the addition of any particular substrate). Eg. Pyrophosphatase.

2. Inducible Enzymes

It is present only in trace amounts or not at all but quickly increases in concentration when its substrate is present. Eg. Amidase.

Origin and state of soil enzymes

Although the general origins of soil enzymes are (a) microorganisms-living and dead, (b) plant roots and plant residues and, (c) soil animals; the state of soil enzymes in the soil is different as below. In soil the enzymes are stabilized in clay,organic matter, and also with organic matter clay complexes, which offers more advantages for the soil enzymes. Most activities are associated with clays. Due to stabilization in clays, the enzymes get increased resistance to proteolysis and microbial attack as well as the inactivation temperature gets increased.

Soil enzymes as indicators of soil health

Soil health can be defined as the continued capacity of soil to function as a vital living system, within the ecosystem and land-use boundaries, to sustain biological productivity, promote the quality of air and water environments, and maintain the plant, animal, and human health. Enzymes are the direct mediators for biological catabolism of soil organic and mineral components. Thus, these catalysts provide a meaningful assessment of reaction rates for important soil processes.Soil enzyme activities (1) are often closely related to soil organic matter, soil physical properties and microbial activity or biomass, (2) changes much sooner than other parameters, thus providing early indications of changes in soil health, and (3) involve simple procedures. Besides, soil enzyme activities can be used as measures of microbial activity, soil productivity, and inhibiting effects of pollutants.

Importance of soil enzymes

Soil enzymes are important in soil functioning because of the following features:

a. they play a critical role in the decomposition of organic materials and the transformation of organic matter

b. they release available nutrients to plants

c. they participate in N_2 fixation, nitrification, and denitrification processes

d. they take part in the detoxification of xenobiotics, such as pesticides, industrial wastes, etc.,

Role of Soil Enzymes

Enzyme	Organic Matter Substances Acted On	End Product	Significance	Predictor of Soil Function
β- glucosidase	carbon compounds	glucose (sugar)	energy for microorganisms	Organic matter decomposition
Amidase	carbon and nitrogen compounds	ammonium (NH_4)	plant-available NH_4	Nutrient cycling

contd...

Urease	nitrogen (urea)	ammonia (NH_3) and carbon dioxide (CO_2)	plant-available NH_4	Nutrient cycling
Phosphatase	phosphorus	phosphate (PO_4)	plant-available P	Nutrient cycling
Sulfatase	sulfur	sulfate (SO_4)	plant-available S	Nutrient cycling

1. Amylase

Amylase is a starch hydrolyzing enzyme. It is known to be constituted by α-amylase and β-amylase. The α- amylases are synthesized by plants, animals, and micro-organisms, whereas, β-amylase is mainly synthesized by plants. This enzyme is widely distributed in plants and soils so it plays a significant role in the breakdown of starch. Normally α-amylase converts starch like substrates to glucose and/or oligosaccharides and β-amylase converts starch to maltose.

2. Arylsulphatase

The relative proportions of inorganic and organic S forms in soils vary widely. However, the organic S compounds are dominant and a large proportion of the organic S is present as ester sulfates. Although ester sulfates are considered to be the most labile form of soil organic S, they are unavailable to plants and must be hydrolyzed to inorganic SO_4^{2-} before plant uptake. Arylsulfatase is involved by cleaving the O - S bond and is believed to make a major contribution to the mineralization of ester sulfate in soils. They are responsible for the hydrolysis of sulphate esters in the soil and are secreted by various bacteria into the external environment as a response to sulphur limitation. Its occurrence in different soil systems is often correlated with microbial biomass and the rate of S immobilization.

3. Cellulases

Cellulases are a group of enzymes that catalyze the degradation of cellulose; polysaccharides build-up of β -1, 4 linked glucose units. Cellulases in soils are derived mainly from plant debris incorporated into the soil, and that a limited amount may also originate from fungi and bacteria in soils. The cellulases system comprises of three major types of enzymes. They include endo-1, 4 - β -glucanase which attacks the cellulose chains at random, exo -1, 4 - β - glucanase (Cellobiohydralse) which removes glucose or cellobiose from the non-reducing end of the cellulose chains, and β-glucosidase which hydrolyzes cellobiose and other water-soluble cellodextrins to glucose.

Action of cellulase enzyme complex

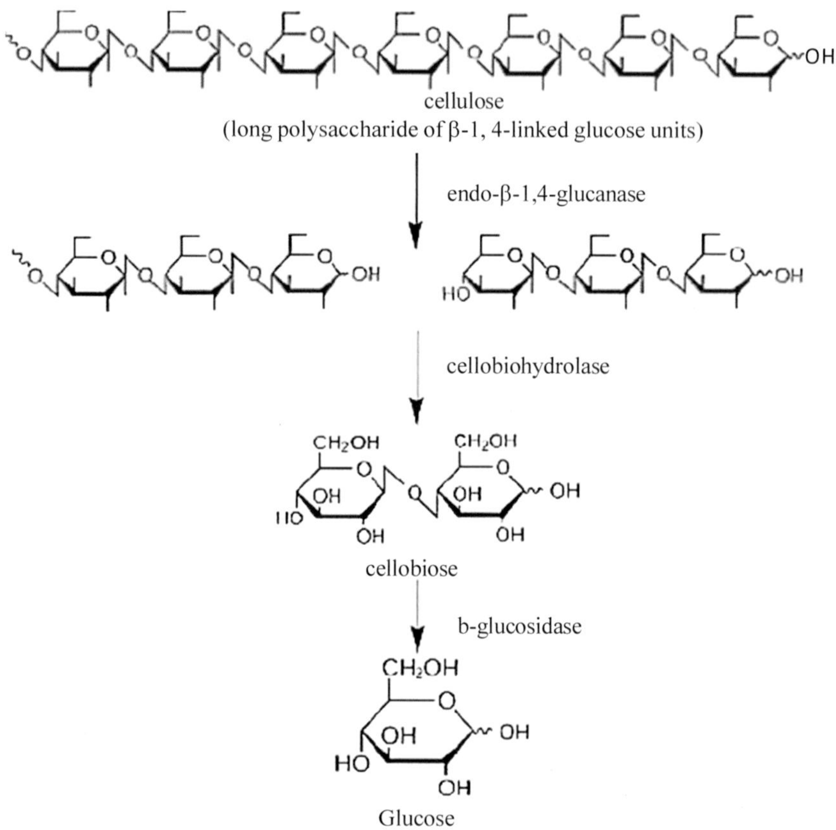

4. β-Glucosidase

The enzyme β-Glucosidase is involved in the enzymatic degradation of cellulose, the main component of plant polysaccharides. Cellulose consists of polymer chains of β-1,4, linked glucose units and its enzymatic degradation initiated by endo β -1, 4- glucanase, which breaks cellulose chains into smaller units, and cellobiohydrolase, which cleaves the dimer cellobiose (two β -1,4 linked glucose units) from the reducing 'ends' of the molecules. β-Glucosidase completes the hydrolysis process by catalyzing the cleavage of cellobiose to release two moles of glucose per mole of cellobiose and, therefore, regulates the supply of an important energy source for microorganisms. β-Glucosidase is derived mostly from soil heterotrophic bacteria and few members of fungi.

5. Chitinase

Chitinases are enzymes that hydrolyze the chitins to it monomers by breaking the glycosidic bonds. Chitinases are present widely in various organisms like viruses, plants, animals, fungi, bacteria, insects and it plays a diverse role in these organisms. Chitinases are divided into 2 broad categories: Endo-chitinases and Exo-chitinases. Endo-chitinases cleaves randomly at the internal sites of chitin and produces low molecular weight multimers such as chitotriose and chitotetraose. Exo-chitinases are further classified into 2 subcategories: chitobiosidases and 1-4-β glucosaminidases. The release of diacetylchitobiose which starts at the non-reducing end of chitin microfibril is zcatalyzed by chitobiosidases. The oligomers obtained by endo-chitinases and chitobiosidases are cleaved by β -(1, 4) glucosaminidases. Chitinases are involved in many physiological and bioconversion processes and they play aa major role in nutrition and parasitism in bacteria.

Chitin polymer and mechanism of action of chitinase enzymes

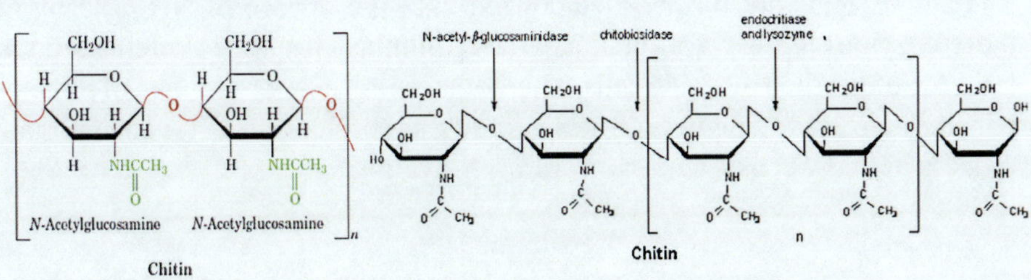

6. Dehydrogenases

Among all enzymes in the soil environment, dehydrogenases are one of the most important and are used as an indicator of overall soil microbial activity, because they occur intracellularly in all living microbial cells. They are tightly linked with microbial oxidoreduction processes and they do not accumulate extracellularly in the soil. Dehydrogenases play a significant role in the biological oxidation of soil organic matter by transferring hydrogen from organic substrates to inorganic acceptors. Active dehydrogenases can utilize both O_2 and other compounds as terminal electron acceptors, although anaerobic microorganisms produce most dehydrogenases. Dehydrogenase activity reflects the metabolic ability of the soil and its activity is considered to be proportional to the biomass of the microorganisms in the soil.

7. Phosphatases

Phosphatases are a broad group of enzymes that are capable of catalyzing the hydrolysis of esters and anhydrides of phosphoric acid. In soil ecosystems, these enzymes are believed to play critical roles in P cycles. Apart from being good indicators of soil fertility, phosphatase enzymes play key roles in the soil system. There are different types of phosphatase in soil. They are phosphomonoesterases (acid and alkaline phosphatase), phosphodiesterases, phosphotriesterases, Adenosine triphosphatase, and pyrophosphatases. Soil phosphatases are extracellular and it is secreted from gram-positive bacteria and algae.

8. Proteases

Proteases are ubiquitous and originate from several different sources in soil, including microorganisms *viz.*, bacteria, actinobacteria, fungi, plants, and animal excrements. Protein that is added to soil is readily decomposed by proteases and peptidases into smaller, membrane-permeable peptides and amino acids. The amino acids are further metabolized with the release of NH_4^+. Many of these enzymes are extracellular, as a large number of native proteins are too large to be absorbed by living cells. Microorganisms cause breakdown of proteins with the help of proteolytic enzymes is known as "proteolysis". It is an important process in many ecosystems concerning N-cycling.

9. Urease

Soil urease originates mainly from plants and micro-organisms found as both intracellular and extracellular enzymes. Urease is responsible for the hydrolysis of urea fertilizer applied to the soil into NH_3 and CO_2 with a parallel rise in soil pH. In turn, it results in a rapid N loss to the atmosphere through NH_3 volatilization. Urease can be produced by bacteria, yeasts, fungi, and algae, as well as plants. Urease may be synthesized constitutively in some organisms, but most often urease expression is under N regulation. The enzyme synthesis is inhibited when cells grow in the presence of a preferred N source such as NH_4^+. In contrast, urease production is activated in the presence of urea or alternative N sources.

Applications of soil enzymes

- Correlation with soil fertility
- Correlation with microbial activity
- Correlation with biochemical cycling of various elements in soil (C, N, S)
- Degree of pollution (heavy metals, SO_4)

- To assess the successional stages of an ecosystem
- Forensic purposes
- Rapid degradation of pesticides
- Disease studies
- Enzyme activity in soil fluctuates with the environment

Factors affecting enzymes activities in soil

Numerous factors can influence enzyme activity in the soil. Natural parameters (e.g., seasonal changes, geographic location, in situ distribution, physical-chemical properties, the content of organic matter, and clay) usually affect the enzyme activity level by influencing both the production of enzymes by plants and microorganisms and their persistence under natural conditions. The physical and chemical properties of soil are involved in the immobilization and stabilization processes of most extracellular enzymes. High content of clay or humus colloids is usually associated with stable but less active enzymes. Agricultural activities and environmental pollution (e.g., fertilizers, pesticides, tillage, heavy metals, PAHs) may affect the chemical composition and structural characteristics of soil, which in turn will influence the abundance of soil microorganisms, the enhancement or suppression of enzyme production and the overall activity of an enzyme in soil.

pH and temperature

Soil enzymes have varying optimum pH and temperature values at which they function most effectively. For example, the activity of phosphatase, arylsulfatase, and amidase involved in phosphorus, sulfur, and nitrogen cycling, respectively, is strongly correlated to variations in soil pH. Since enzyme structure and substrate binding can be altered by heat and extreme cold temperature, enzyme activity decreases above and below the optimum temperature.

Moisture

The activity of many enzymes often correlates with soil moisture content, as well. Drought may suppress enzyme activity.

Soil texture

Soil texture influences enzyme activity, and normally enzyme activities are significantly and positively correlated with clay content. Clayey soils have a greater ability to store organic matter that promotes microbial communities, and clay forms clay-enzyme complexes. In contrast, sandy soils tend to exhibit low rates of enzyme activity because they are naturally low in organic matter and have poor water holding capacity which results in lower microbial biomass and therefore lower enzyme activity.

Plant roots

Plant roots stimulate enzyme activity because of their positive effect on microbial activity and the production of exudates rich in substrates acted on by enzymes.

End product inhibition

The end products of enzymatic reactions can inhibit enzyme activity by feedback inhibition. For example, phosphatase activity increases in phosphorus-deficient soil, but its activity decreases in soil with high phosphorus concentration. Similarly, urease activity may be suppressed by ammonia-based nitrogen fertilizer because ammonium is the end product of urease activity.

Compaction

Compaction may limit the activity of enzymes involved in nutrient mineralization because of decreased oxygen in the soil. On the other hand, anaerobic conditions from compaction or water saturation increase enzymatic reaction rates related to denitrification.

Chemicals and pollutants in soil

Application of materials containing heavy metals can reduce enzyme activity (e.g., amidase)

Role in soil fertility

Soil fertility depends on the physical, chemical, and biological properties of the soil. Among the many biological properties that have potential as sensitive indicators of soil quality and fertility, enzyme activities often provide a unique integrative biological assessment of soil function, especially those that catalyze a wide range of soil biological processes, such as dehydrogenase, urease, and phosphatase. Soil enzymes play key biochemical functions in the overall process of the transformation of organic matter and soil nutrient cycling in the soil system.

8

Bio Geochemical Cycling

All organisms require the six elements CHNOPS in sizable quantities, and that these minerals may occur in different forms. Depending on whether the local environment is aerobic or anaerobic, the exact chemical form of any of these elements may change quickly and drastically. Example: If the soil becomes waterlogged and anaerobic, sulfate will be converted into hydrogen sulfide and nitrate into nitrogen gas or ammonia (*via* anaerobic respiration). As the water recedes and the soil becomes aerobic again, the ammonia is oxidized back to nitrate in stages. All of this (and many other processes) occur largely because of the activities of soil microbes. Without these microbes, life as we know it would not occur because necessary elements would remain tied up in unusable chemical forms and gradually be removed by sedimentation. To maintain the level of these components it is essential that they undergo a regular process of recycling. Different microorganisms bring about this process of recycling through various transformations.

Carbon Cycle

The most important element in the biological realm and substance that serve as the cornerstone of the cell structure is carbon. It constituents about 40-50% of all living organisms, yet the ultimate source is the CO_2 that exists in a perennially short supply, only 0.03% of the earth's atmosphere, which undergoes a cyclic change from an oxidized to reduced state. The concentration of carbon in living matter (18%) is almost 100 times greater than its concentration in the earth (0.19%). Hence, living things extract carbon from their non-living environment. For life to continue, this carbon must be recycled. Carbon exists in the nonliving environment as carbon dioxide (CO_2) in the atmosphere and dissolved in water (forming HCO_3") carbonate rocks (limestone and coral = $CaCO_3$) deposits of coal, petroleum, and natural gas derived from once-living things dead organic matter, e.g., humus in the soil. Carbon enters the biotic world through the action of autotrophs: primarily photoautotrophs, like plants and algae, that use the energy of light to convert carbon dioxide to organic matter through photosynthesis and to a small extent, chemoautotrophs - bacteria and Archaeans that do the same but use the energy derived from the oxidation of molecules in their substrate.

The carbon cycle revolves about CO_2 and its fixation and regeneration. It is estimated that $1.3x10^{14}$ kg CO_2 is fixed annually in the biosphere.The green plants utilize CO_2 as their sole carbon source, and the carbonaceous matter synthesized serves to supply carbon to other heterotrophic organisms and animals. Upon the death of plants and animals, microbes assume a dominant role in the carbon cycle. The dead tissues are degraded and transformed into microbial cells and humus or soil organic fraction. Further decomposition of these materials leads to the production of CO_2 (producing CO_2 if oxygen is present, methane - CH_4 if oxygen is not present), and once again it is recycled.

The uptake and return of CO_2 are not in balance. The carbon dioxide content of the atmosphere is gradually and steadily increasing. This increase is surely "anthropogenic" that is, caused by human activities: burning fossil fuels (coal, oil, natural gas), clearing and burning of forests, especially in the tropics. In recent decades, large areas of the Amazon rain forest have been cleared for agriculture and cattle grazing.

Carbon Assimilation

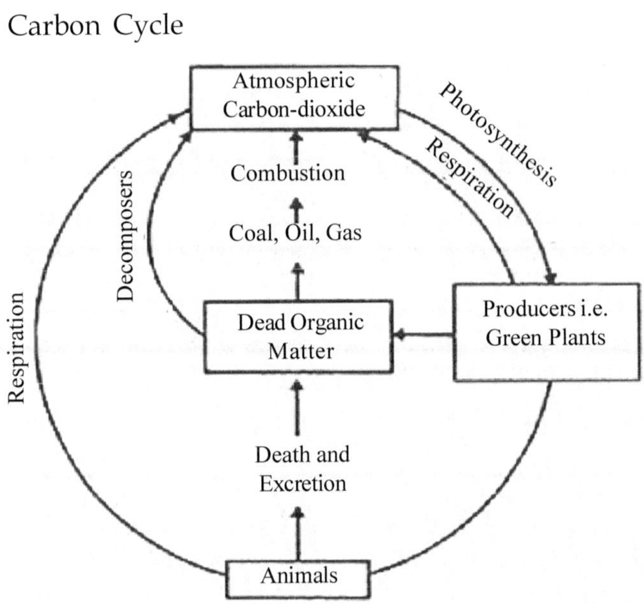

The process of converting substrate to protoplasmic carbon is known as assimilation. Under aerobic conditions 20-40% of the substrate carbon is assimilated, the remainder is released as CO_2. Fungi are more efficient, in their metabolism, since they convert carbon into cell carbon as filaments and releases less of CO_2. Around 30-40% is used to form new mycelium during the decomposition. Compared to fungi, bacteria are less efficient.

Mineralization

The conversion of organic carbon substances into the inorganic form of carbon is known as carbon mineralization.

$$\text{Organic} \rightarrow \text{Inorganic}$$

Immobilization

The assimilation of nutrients by the microorganisms called immobilization, which reduces the quantity of available nutrients to the plants.

$$\text{Inorganic} \rightarrow \text{Organic}$$

9

Decomposition of Organic Matter and Humus Formation

Soil Organic Matter

Soil organic matter plays an important role in the maintenance and improvement of soil properties. It is a dynamic material and is one of the major sources of nutrient elements for plants. Soil organic matter is derived to a large extent from residues and remains of the plants together with the small quantities of animal remains excreta, and microbial tissues. Soil organic matter is composed of three major components i.e. plant residues, animal remains and dead remains of microorganisms. Soil organic matter comprises residues of plants and animals and these compounds occur in soil in close combination with inorganic substances. Animals and plant residues are made up of complex carbohydrates, simple sugars, starch, cellulose, hemicellulose, pectins, gums, mucilage, proteins fats, oils, waxes, resins, alcohols, aldehydes, ketones, organic acids, lignin, phenols, tannins, hydrocarbons, alkaloids, pigments, etc.

The organic matter subjected to microbial decay in the soil comes from several sources like plant remains, forest litter, incorporation of plant tissues, animal tissues, and excretory products. The chemistry of organic matter is very complex, and investigations of the transformations and the responsible organisms have therefore been extremely interesting. The organic constituents of the plants are commonly divided into six categories.

a. **Cellulose** - Most abundant 15-60% of the dry weight

b. **Hemicellulose** - 10-30% of the plant dry weight

c. **Lignin** - 5 - 30% of the plant dry weight

d. **Water-soluble fraction** - 5 - 30%, includes simple sugars and amino acids

e. **Ether and alcohol-soluble constituents**, a fraction containing fat, oils, waxes, resins, and pigments

f. **Proteins** - As the plant ages, the content of water-soluble constituents, proteins, and minerals decreases.

Hence the percentage of the abundance of cellulose, hemicellulose, and lignin rises. The relationship between organic matter and plant growth may be direct or indirect.

- Organic matter is a natural substrate for saprophytic microorganisms and provides nutrition to plants indirectly through the activity of soil microorganisms.
- It is essential for the formation of soil aggregates and hence soil structure which ultimately determines the soil aeration and rooting habit of plants
- Organic matter helps in the conservation of soil nutrients by preventing erosion and surface runoff of nutrients.

Organic Matter (Undecomposed)

A. Organic

- **Nitrogenous**
  1. Water-soluble eg. Nitrates, ammonical compounds, amides, amino acids, etc.
  2. Insoluble eg. proteins, nucleoproteins, peptides, alkaloids, purines, pyridines, chitin, etc.
- **Non-Nitrogenous**
  1. Carbohydrates eg. Sugars, starch, hemicellulose, gums, mucilage, pectins, etc.
  2. Miscellaneous: eg. Lignin, tannins, organic acid, etc.
  3. Ether soluble: eg. Fats, oils, wax, etc.

B. Inorganic

The organic matter in the soil is, therefore made up of a large number of substances of widely different chemical composition and the amount of each substance varies with the type, nature, and age of plants. For example cellulose in a young plant is only half of the mature plants; water-soluble organic substances in young plants are nearly double to that of older plants. Among the plant residues, leguminous plants are rich in proteins than the non-leguminous plants. Grasses and cereal straws contain a greater amount of cellulose, lignin, hemicelluloses than the legumes and as the plant gets older the proportion of cellulose hemicelluloses, and lignin gets increased. Plant residues contain 15-60% cellulose 10-30 % hemicellulose, 5-30% lignin, 2-15 % protein, and 10% sugars, amino acids, and organic acids. These differences in the composition of various plant and animal residues have great significance on the rate of organic matter decomposition in general and of nitrification and humification (humus formation) in particular.

Organic matter decomposition (Aerobic)

The soil microorganisms play an important role in the decomposition of organic matter. Bacteria are the dominant group in the decomposition; mostly heterotrophic organisms (use energy from organic sources such as sugars, starch, cellulose, and protein) are actively involved. Autotrophic organisms occupy a small portion of the biomass in the soil are not directly involved in organic matter decomposition. Soil fungi are mostly heterotrophs and use organic residues easily.

Soil algae contribute a small amount of organic matter through their biomass, but they do not have an active role in organic matter decomposition.

Genera of microorganisms capable of utilizing different components of organic matter as reported by several workers: Groups of microbes: F-Fungi; B-Bacteria; A-Actinobacteria

Nature of substrate in organic matter	Group	Genera of microorganisms
Cellulose	Fungi	*Alternaria, Aspergillus, Chaetomium, Coprinus, Fomes, Fusarium, Myrothecium, Penicillum, Polyporus, Rhizoctonia, Rhizopus, Trametes, Trichoderma, Trichothecium, Verticillium, Zygorynchus*
	Bacteria	*Achrombacter, Angiococcus, Bacillus, Cellfalcicula, Cellulomonas, Cellvibrio, Clostrium, Cytophaga, Polyangium, Pseudomonas, Sorangium, Sporocytophaga, Vibrio*
	Actinobacteria	*Micromonospora, Nocardia, Streptomyces, Streptosporangium*
Hemicellulose	Fungi	*Alternaria, Fusarium, Trichothecium, Aspergillus, Rhizopous, Zygorynchus, Chateomium, Helminthosporium, Penicillium, Coriolus, Fomes, Polyporus*
	Bacteria	*Bacillus, Achromobacter, Pseudomonas, Cytophaga, Sporocytophaga, Lactobacillus, Vibrio*
	Actinobacteria	*Streptomyces*
Lignin	Fungi	*Clavaria, Clitocybe, Collybia, Flammula Hypholoma, Lepiota, Mycena, Pholiota, Arthrobotrys, Cephalosporium, Humicola*
	Bacteria	*Pseudomonas, Flavobacterium*
Pectin	Fungi	Several fungi
	Bacteria	*Bacillus, Clostridium, Pseudomonas, Erwinia,*
	Actinobacteria	*Streptomyces*

Process of decomposition

When plant and animal residues are added to the soil, the various constituents of the soil organic matter are decomposed simultaneously by the activity of microorganisms, and carbon is released as CO_2, and nitrogen as $NH_4 \rightarrow NO_3$ for the use by plants. Other nutrients are also converted into plant usable forms. This process of release of nutrients from organic matter is called mineralization. The insoluble plant residues constitute the part of humus and soil organic matter complex. The final product of aerobic decomposition is CO_2 and that of anaerobic decomposition is hydrogen, ethyl alcohol, methane, various organic acids, and carbon dioxide. Soil organisms use organic matter as a source of energy and food.

The process of decomposition is initially fast, but slows down considerably as the supply of readily decomposable organic matter gets exhausted. Sugars, water-soluble nitrogenous compounds, amino acids, lipids, starches, and some of the hemicelluloses are decomposed first at a rapid rate, while insoluble compounds such as cellulose, hemicellulose, lignin, proteins, etc. which forms the major portion of organic matter are decomposed later slowly. Thus, the organic matter added to the soil is converted by oxidative decomposition to simpler substances which are made available in stages for plant growth and the residue is transformed into humus.

During the decomposition of organic matter three separate simultaneous processes can be distinguished. The important changes during decomposition are:

- Plant and animal tissue constituents disappear under the influence of enzymes.
- Synthesis of new microbial cells so that proteins, polysaccharides, and nucleic acids of bacteria and fungi appear.
- Third, certain end products of the breakdown are excreted into the surroundings to accumulate or to be further metabolized.

Anaerobic decomposition

The main products of aerobic carbon mineralization are CO_2, water, cells, and humus components. In the absence of O_2 organic carbon is incompletely metabolized, intermediary substances accumulate, and abundant quantities of CH_4 and smaller amounts of H_2 are evolved. Energy yield during anaerobic fermentation is low, resulting in fewer microbial cells per unit of organic carbon degraded. Consequently, organic matter breakdown is consistently slower under total anaerobiosis than in environments containing adequate O_2. The rate in waterlogged soils is intermediate between the two extremes.

Flow or organic matter decomposition

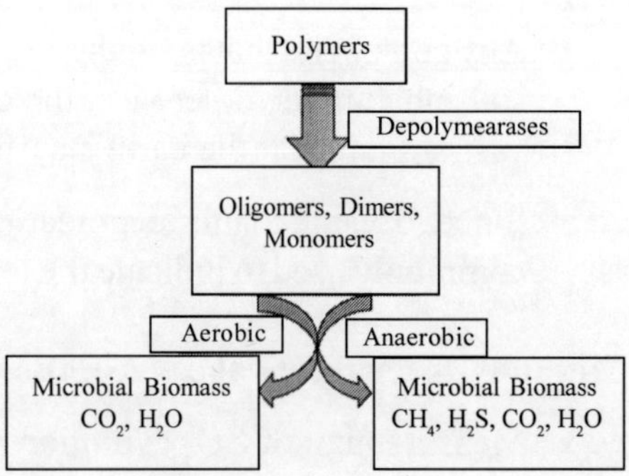

When soil is waterlogged or flooded there is a shift from aerobic to anaerobic transformation. Formation and accumulation of organic acids *viz.,* acetic, formic, butyric, lactic, and succinic acids appear too, these are frequently detrimental to root development. Organic acids accumulate because of the fermentative character of the microflora of wet soils. The anaerobic carbon transformation is thus characterized by the formation of organic acids, alcohols, $CH_{4,}$ and CO_2 as major end products.

Under anaerobic conditions, decomposition of organic residues takes place by the activity of both mesophilic, thermophilic microorganisms resulting in the production of CO_2, H, ethyl alcohol, and organic acids. Among mesophiles, bacteria are more active than fungi or actinobacteria in the cellulolytic activity. They belong to genus *Clostridium* which is numerous in manure pits but rarely encountered in cultivated arable soils. In compost pits both mesophilic and thermophilic (bacterial and actinobacteria) are important in the breakdown of cellulose substances.

The primary microbial colonizers initially breakdown the complex CH_2O and proteins into organic acids and alcohols. At a later stage, the methane bacteria, which are strict anaerobes, begin to act upon the secondary substrate chiefly lactic and butyric acids and ferment them into CH_4 and CO_2.

Organic matter decomposition serves two important functions

a) Provide energy for growth

b) Supply carbon for the formation of new cell materials

End products

The end products of decomposition are CO_2, H_2O NO_3, SO_4, CH_4, $NH_{4,}$ and H_2S depending on the availability of oxygen.

Importance of organic matter decomposition

- The important function is the breakdown of organic matter by which CO_2 available for photosynthesis is replenished
- Any compound that is synthesized biologically is subject to destruction by soil inhabitants, otherwise, the compounds would have accumulated in vast amounts on the earth's surface
- Since, organic matter degradation is a property of all heterotrophs, it is commonly used to indicate the level of microbial activity.

Methods to evaluate the decomposition rate

- Measurement of CO_2 evolution or O_2 uptake
- Determination of decrease in the organic matter either chemically or by weight loss
- Observations on the disappearance of specific constituents such as cellulose, hemicellulose, or lignin.

Microbiology of decomposition of various constituents in organic matter

The microbiology of decomposition/degradation of some of the major constituents (*viz*. Cellulose, Hemicellulose, Lignin, Proteins etc.) of soil organic matter/plant residues are discussed in brief in the following paragraphs.

a) Decomposition of cellulose

Cellulose is the most abundant carbohydrate present in plant residues/organic matter in nature. It is a CH_2O composed of glucose units bound by β-linkage at carbon 1 and 4 of the sugar molecule. The cellulose concentration of higher plants is never fixed. It is a polymer of glucose and abundant organic material in nature changes with age and type of plants. Woody materials have more cellulose and succulent tissues have poor, but the concentration increases as the plant mature. Cellulose breakdown in the soil is influenced by several environmental factors.

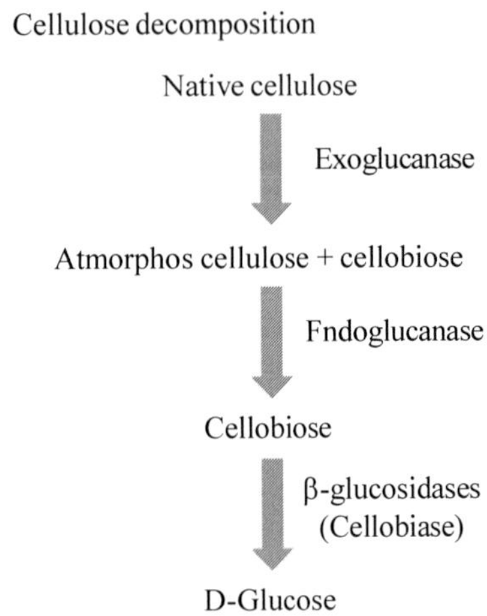

Aerobic organisms convert the cellulose into 2 major products: CO_2 + cell substance, but certain groups release a small amount of organic acids. It is however resistant to microbial decomposition. When cellulose is associated with pentosans (xylan, mannans) it undergoes rapid decomposition. When associated with lignin, the decomposition rate is very low. Degradation is carried out by cellulases enzyme complex that converts cellulose into glucose.

i. Exoglucanase

ii. Endoglucanase

iii. β -glucosidase

i. Cellobiohydrolases (CBHs) are exoglucanases that cleave disaccharide units from the reducing or non-reducing end of the cellulose molecule.

ii. Endoglucanases (EGs) cleave bonds in the amorphous region of cellulose, generating ends that act as CBH substrates.

iii. β -glucosidase (cellobiases)cleaves the disaccharides into glucose monomers.

The decomposition of cellulose occurs in two stages: (i) in the first stage the long chain of cellulose is broken down into cellobiose and then into glucose by the process of hydrolysis in the presence of enzymes cellulase and cellobiase, and (ii) in second stage glucose is oxidized and converted CO_2 and water.

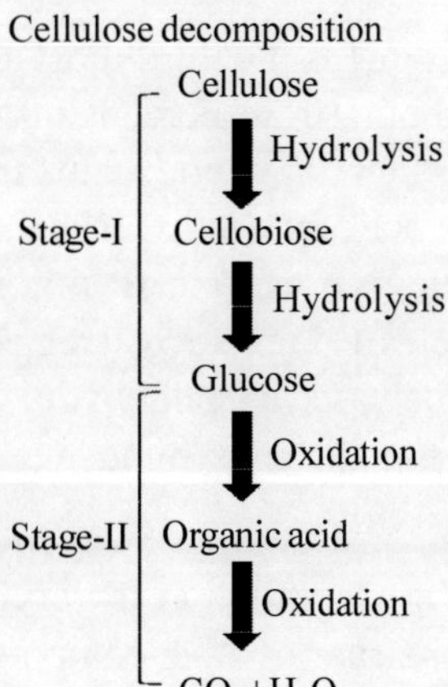

The intermediate products formed/released during enzymatic hydrolysis of cellulose (eg. cellobiose and glucose) are utilized by the cellulose-decomposing organisms or by other organisms as a source of energy for biosynthetic processes. The cellulolytic microorganisms responsible for the degradation of cellulose through the excretion of enzymes (cellulase and cellobiase) are fungi, bacteria, and actinobacteria.

Most cellulolytic bacteria do not excrete significant amounts of cellulase but fungi are found to excrete these enzymes. The soluble sugars released by enzymatic hydrolysis are later utilized by the same or other microorganisms for biosynthetic purposes.

b) Decomposition of hemicelluloses

Hemicelluloses are water-soluble polysaccharides and consist of hexoses, pentoses, and uronic acids and are the major plant constituents second only in the quantity of cellulose, and sources of energy and nutrients for soil microflora. It may be either homo or heteropolymers. When they are added to the soil, degradation takes place at a faster rate in initial stages. The hemicelluloses such as mannans are decomposed rapidly while galactons (a polymer of galactose) are decomposed slowly. Many soil microorganisms utilize hemicellulose in both aerobic as well as anaerobic conditions. The microbial degradation occurs through the agency of extracellular enzymes called hemicellulases.

On hydrolysis hemicelluloses are converted into soluble monosaccharide/sugars (eg. xylose, arabinose, galactose, and mannose) which are further converted to organic acids, alcohols, CO_2 and H_2O and uronic acids are broken down to pentoses and CO_2. Various microorganisms including fungi, bacteria, and actinobacteria both aerobic and anaerobic are involved in the decomposition of hemicelluloses.

c) Decomposition of Lignin

Lignin is the third most abundant constituent of plant tissues, and accounts for about 10-30 percent of the dry matter of mature plant materials. The lignin content of young plants is low and gradually increases as the plant grows old. It is one of the most resistant organic substances for the microorganisms to degrade however certain Basidiomycetous fungi are known to degrade lignin at slow rates. Complete oxidation of lignin results in the formation of aromatic compounds such as syringaldehyde, vanillin, and ferulic acid. The final cleavages of these aromatic compounds yield organic acids, carbon dioxide, methane, and water.

It consists of heterocyclic aromatic organic molecules containing C, H, and O. The degradation is very slow and the rate of decomposition depends on the presence of other compounds such as cellulose and hemicellulose acid. Lignin is highly resistant to microbial degradation. Degradation is a complex process as follows.

Lignin degradation

Lignin → Coniferyl ether → Coniferyl alcohol → Coniferyl aldehyde → Vanillin → Vannillic acid → Protocatechuic acid → ring cleavage → Organic acids, CO_2, CH_4 & H_2O

Brown rot fungi, White rot fungi, and soft rot fungi are mainly involved in the degradation of lignin. Brown rot fungi degrade cellulosic component and white rot degrade the lignin component of lignocelluloses. Ligninases, peroxidases, and polyphenol oxidases are the enzymes involved in the degradation of lignin. *Phanerochaete chrysosporium* and *Chaetomium* are lignin-degrading fungi.

d) Decomposition of pectin

Pectic substances are polysaccharides found in the constituents of the middle lamella and the primary and secondary cell walls of plants. They are made up of galacturonic acid units bound in a long chain. There are three types of pectic substances, *viz.*, pectin, protopectin, and pectic acid. These compounds, though, closely related require specific enzymes to get hydrolyzed. The enzymes involved in pectin digestion are collectively known as pectinases.

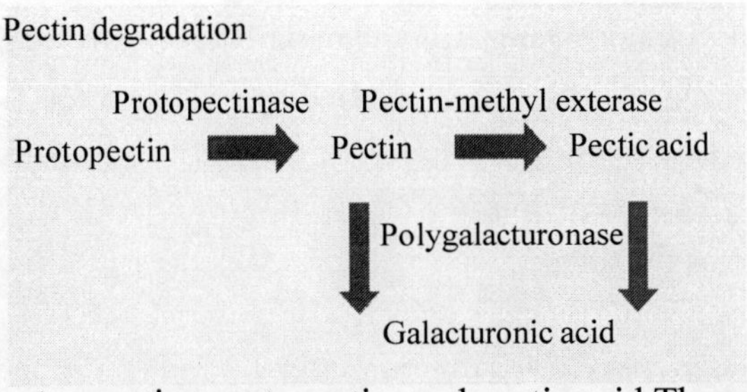

e) Decomposition of Starch

Starch or amylum is a carbohydrate consisting of a large number of glucose units joined by glycosidic bonds. They are carbohydrate reserves in plants. Starch consists of a mixture of two types of polymers: amylose and amylopectin.

Amylose is linear polysaccharide-linked by α-(1- 4) glycosidic linkage between glucose residues and it is water-soluble.

Amylopectin is branched linked by α (1-4) between glucose residues and also by α-(1-6) linkage and it is water-insoluble.

Starch molecules are broken down by an enzyme known as amylase. Bacteria and fungi secrete amylases for extracellular digestion. When they have broken down the insoluble starch, the soluble end products such as (glucose or maltose) are absorbed into their cells. Many microorganisms produce this amylase enzyme, the most common organisms are *Bacillus subtilis, B. licheniformis, B. amyloliquifaciens* and *Aspergillus niger*.

Microbial amylases usually contain four types.

1. α-Amylase

It is produced by many bacteria and fungi. It is acting at a random location along the starch chain. α - Amylase breaks down long-chain carbohydrates ultimately yielding maltotriose and maltose or glucose. It is called endoglucanase because it normally attacks internal α -1,4 bonds.

2. β - Amylase

It is found primarily in plants and frequently made by bacteria. It removes maltose molecules from the nonreducing ends of starch. β-amylase catalyzes the hydrolysis of the second α-1-4 glycosidic bond, cleaving off two glucose units (maltose) at a time. As it removes molecules from the external end, it is called an exoglucanase.

3. γ-Amylase (Glucoamylase)

γ-Amylase will cleave α (1-6) glycosidic linkages, as well as the last α (1-4) glycosidic linkages at the nonreducing end of amylose and amylopectin, yielding glucose.

4. Pullulanase or isoamylase

A special kind of glucanase, an amylolytic enzyme that degrades pullulan. Pullulan is a polysaccharide polymer consisting of maltotriose units. Pullulan is produced from starch by a fungus *Auerobasidium pullulans*.

f) Decomposition of protein

Proteins are complex organic substances containing nitrogen, sulphur, and sometimes phosphorus in addition to carbon, hydrogen, and oxygen. During decomposition of organic matter, proteins are first hydrolyzed to several intermediate products eg. Proteases, peptides, etc. collectively known as polypeptides.

The intermediate products so formed are then hydrolyzed and broken down ultimately to individual amino acids, or ammonia and amides. The process of hydrolysis of proteins to amino acids is known as “aminization or ammonification”, which is brought about by certain enzymes, collectively known as “proteases” or “proteolytic” enzymes secreted by various microorganisms. Amino acids and amines are further decomposed and converted into ammonia. During ammonification, various organic acids, alcohols, aldehydes, etc. are produced which are further decomposed finally to produce carbon dioxide and water.

All types of bacteria, fungi, and actinobacteria can bring about decomposition of proteins. In acid soils, fungi are predominant, while in neutral and alkaline soil bacteria are dominant decomposers of proteins.

Humus

A dark-colored and fairly stable soil organic matter with known and unknown physical and chemical properties is called humus. It is an integral part of the organic matter complex in soil. Humus can be defined as ligno-protein complex containing approximately 45 % lignin compounds, 35 % amino acids, 11 % carbohydrates, 4 % cellulose, 7 % Hemicellulose, 3 % fats, wax, resins and 6 % other miscellaneous substances, including plant growth substances and inhibitors.

The age and composition of the humus are dependent on its origin and environment. Bacterial and algal protoplasm contribute a good deal to the nutritive value of humus. Soil microorganisms take part in humus formation. Some fungi such as *Penicillium*, *Aspergillus,* and actinobacteria produce dark humus like substances, which serve as structural units for the synthesis of humic substances. The organic fraction of soil is often termed as humus. It is a product of synthetic and decomposing activities of the microflora. Since it contains the organic C and N needed for microbial development, it is the dominant food reservoir.

Humus formation

Once the plant or animal remains fall on are incorporated into the soil, they are subjected to decomposition. From the original residues, a variety of products are formed. As the original materials and the initial products undergo further decomposition, they are converted to brown or black organic complexes. At this stage any trace of the original remains no longer remains. The native organic fraction originates from two sources: the original plant debris entering the soil and the microorganisms within the soil body. The microorganisms in soil body work upon the former and synthesize microbial protoplasm and new compounds that become part of the organic fraction.

Humus exists in a dynamic state and the chemistry of humus is complex. It has been pointed out that the organic fraction is derived from plant constituents that are modified by the microflora.Constituents of microbial cells and products of microbial metabolism that is relatively resistant to decay and therefore persist for sometime after the death of organisms.

The organic fraction contains compounds of C, H, O, N, P, S, and small amounts of other elements. Only a small portion of the total is soluble in water, but much can be brought into solution by alkali. Humus contains several polymerized substances, aromatic, molecules, polysaccharides, ascorbic acids, polymers of uronic acids, and P containing compounds. No definite composition can be assigned. It should be considered as a portion of the soil that is composed of a heterogeneous group of substances, most having an unknown parentage and an unknown chemical structure. Lignin and lignin-derived molecules have long been considered to be of significance in the formation of humus. It is possible either that simple aromatics released in the microbial attack on lignin polymerize to yield constituents of the soil organic fraction or that partially altered lignin itself give rise to humus constituents. The monomeric portions of humus are similar to the constituents of lignin.

Humus formation from the residues of plant and animal is a two-stage process:

1. The decomposition of the original components of tissues and their conversion by microorganisms into simpler chemical compounds and partially to products of complete mineralization (CO_2, NO_2, NO_3, NH_3, CH_4, H_2O etc.).
2. The synthesis of organic compounds with the formation of high molecular-weight humus substances of specific nature. Micro-organisms participate mainly in the first stage of humus formation the decomposition of the fresh organic materials and the synthetic activity is limited to the re-synthesis of bacterial plasma. In the second stage of humus formation i.e. the condensation reactions (physicochemical reactions) can take place without the participation of micro-organisms. Humus or soil organic matter consists of two major types of compounds-un-humified and humified residues of plant and animal. The un-humified substances are represented by certain organic compounds like, carbohydrates, fats, waxes, and proteins.

On the other hand, the humified fractions are considered as the most active part of humus which consists of a series of highly acidic, yellow to black-colored, high molecular weight polyelectrolytes known as humic acid, fulvic acid and so on. It is generally accepted that humic and fulvic acids are formed by a multiple-stage that includes:

(i) Decomposition of all plant components, including lignin, into simpler monomers;

(ii) Metabolism of the monomers with an accompanying increase in the soil biomass;

(iii) Repeated recycling of the biomass C (and N) with the synthesis of new cells; and

(iv) Con-current polymerization of reactive monomers into high-molecular-weight polymers.

Concepts of Humus Formation

Flaig's concept of humus formation

Lignin, freed of its linkage with cellulose during decomposition of plant residues, is subjected to oxidative splitting with the formation of primary structural units (derivatives of phenylpropane).

1. The side-chains of the lignin-building units are oxidized, demethylation occurs, and the resulting polyphenols are converted to quinones by polyphenoloxidase enzymes.
2. Quinones arising from the lignin (and from other sources) react with N-containing compounds to form dark-colored polymers.

Kononova concept of humus formation

The role of microorganisms as sources of polyphenols has been emphasized byKononova.She concluded that humic substances were being formed by cellulose-decomposing myxobacteria before lignin decomposition.

The stages leading to the formation of humic substances were postulated to be:

1. Fungi attack simple carbohydrates and parts of the protein and cellulose in the medullary rays, cambium, and cortex of plant residues.
2. The cellulose of the xylem is decomposed by aerobic myxobacteria. Polyphenols synthesized by the myxobacteria are oxidized to quinones by polyphenoloxidase enzymes, and the quinones subsequently react with N compounds to form brown humic substances.
3. Lignin is decomposed. Phenols released during decay also serve as source materials for humus synthesis.

Humus formation (Polyphenol theory)

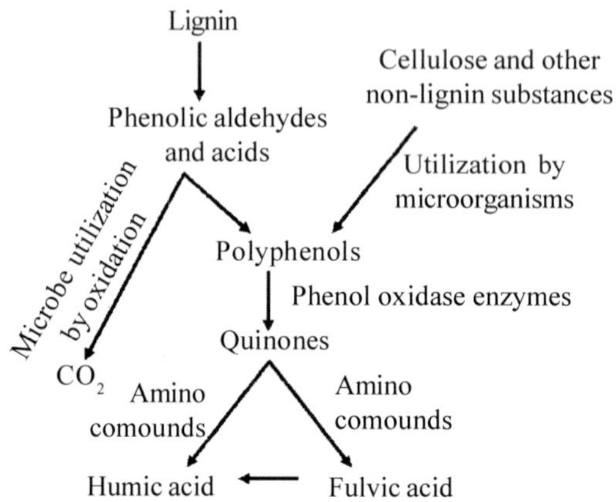

According to present-day concepts, polyphenols derived from plants (e.g. lignin), or synthesized by micro-organisms, are enzymatically converted to quinones, which undergo self-condensation or combine with amino-compounds to form N-containing polymers. An overall scheme for the formation of humus is shown below.

Possible pathways of humus formation

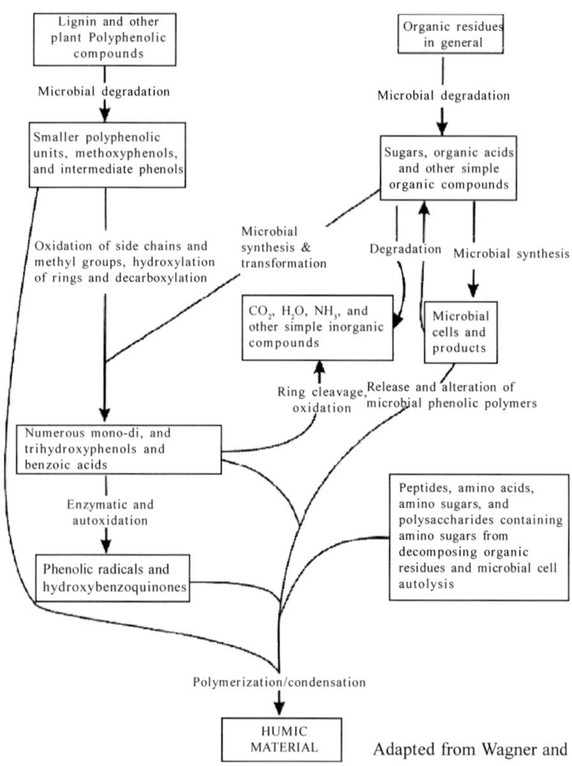

Adapted from Wagner and Wolf, (1998).

Benefits of humic substance

- Improves seed germination, root growth, uptake of minerals by plants, and other physiological effects on plant growth.
- Increases the enzyme activities involved in plant metabolism, since humic acid serves as hydrogen acceptors.
- Increases the cytochrome oxidase activity in root systems results in growth stimulatory effect (on roots)
- It improves the vigor and enhances the yield of plants.
- Humic acid is known to influence the growth and proliferation of microorganisms.
- Humus is a colloidal substance and increases the cation exchange capacity of the soil, hence its ability to store nutrients by chelation.
- Humus can hold the equivalent of 80-90% of its weight in moisture, and therefore increases the soil's capacity to withstand drought.

Factors influencing the rate of organic matter decomposition

In addition to the composition of organic matter, nature and abundance of microorganisms in the soil, the extent of C, N, P, and K., moisture content of the soil and its temperature, pH, aeration, C: N ratio of plant residues and presence/ absence of inhibitory substances (e.g. tannins), etc. are some of the major factors which influence the rate of organic matter decomposition.

As soon as plant and animal residues are added to the soil, there is a rapid increase in the activity of microorganisms. Bacteria are the most abundant organisms playing an important role in the decomposition of organic matter. Majority of bacteria involved in the decomposition of organic matter are heterotrophs and autotrophs are least in proportions which are not directly involved in organic matter decomposition. Actinobacteria and fungi are also found to play an important role in the decomposition of organic matter. Soil algae may contribute a small amount of organic matter through their biomass but they do not have an active role in organic matter decomposition.

a) Aeration

Good aeration is necessary for the proper activity of the microorganisms involved in the decomposition of organic matter. Under anaerobic conditions fungi and actinobacteria are almost suppressed and only a few bacteria *(Clostridium)* take part in anaerobic decomposition. The rate of decomposition is markedly retarded.

It was found that under aerobic conditions 65 percent of the total organic matter decomposes during six months, while under anaerobic conditions only 47 percent organic matter can be decomposed during the same period. Anaerobic decomposition of organic matter results in the production of a large quantity of organic acids and the evolution of gases like methane (CH_4) hydrogen (H_2) and carbon dioxide (CO_2).

b) Temperature

The rate of decomposition is more rapid in the temperature range of 30° to 40°. At temperatures below or above this range, the rate of decomposition is markedly retarded. Appreciable organic matter decomposition occurs at 25° C and further fluctuation in the soil temperature has little effect on decomposition.

c) Moisture

Adequate soil moisture i.e. about 60 to 80 percent of the water-holding capacity of the soil is a must for the proper decomposition of organic matter. Too much moisture leads to insufficient aeration which results in the reduced activity of microorganisms and thereby checks the rate of decomposition.

d) Soil pH/soil reaction

Soil pH affects directly the kind, density, and the activity of fungi, bacteria & actinobacteria involved in the process of decomposition and thereby rate of decomposition of organic matter. The rate of decomposition is more in neutral soils than that of acidic soils. Therefore, the treatment of acid soils with lime can accelerate the rate of organic matter decomposition.

e) C : N ratio

C : N ratio of organic matter has a great influence on the rate of decomposition. Organic matter from diverse plant-tissues varies widely in their C : N ratio (app. 8-10 %). The optimum C : N ratio in the range of 20-25 is ideal for maximum decomposition since a favourable soil environment is created to bring about equilibrium between mineralization and immobilization processes. Thus, low nitrogen content or wide C : N ratio results in the slow decomposition. Protein-rich, young and succulent plant tissues are decomposed more rapidly than protein-poor, mature and hard plant tissues. Therefore, the C : N ratio of organic matter as well as the soil should be narrow/less for better and rapid decomposition.

Thus, high aeration, mesophilic temperature range, optimum moisture, neutral/alkaline soil reaction, and narrow C : N ratio of soil and organic matter are required for rapid and better decomposition of organic matter.

10

Nitrogen Cycle

Soil microorganisms are the most important agents in the cycling or transformation of various elements (N, P, K, S, Fe etc.) in the biosphere; where the essential elements undergo cyclic alterations between the inorganic state as free elements in nature and the combined state in living organisms. Life on earth is dependent on the cycling of nutrient elements from their elemental states to inorganic compounds to organic compounds and back into their elemental states.

The microbes through the process of biochemical reactions convert / breakdown complex organic compounds into simple inorganic compounds and finally into their constituent elements. This process is known as **"Mineralization"**.

Mineralization of organic carbon, nitrogen, phosphorus, sulphur, and iron by soil microorganisms makes these elements available for reuse by plants. In the following paragraphs the cycling or transformations of some of the important elements are discussed.

Nitrogen Cycle

Biological availability of N, P, and K is of considerable economic importance since they are the major plant nutrients derived from the soil. Of the three, N stands out as the most susceptible one to microbial transformations. This element is the key building block of the protein molecule upon which all life is based; it is an indispensable component of the protoplasm of plants, animals, and microorganisms.

Although molecular nitrogen (N_2) is abundant (i.e 78-80 % by volume) in the earth's atmosphere, it is chemically inert and therefore, cannot be utilized by most living organisms and plants. Plants, animals, and most microorganisms depend - on a source of combined or fixed nitrogen (eg. ammonia, nitrate) or organic nitrogen compounds for their nutrition and growth. Plants require fixed nitrogen (ammonia, nitrate) provided by microorganisms, but about 95 to 98 % of soil nitrogen is in organic form (unavailable) which restricts the development of living organisms including plants and microorganisms. Therefore, cycling/ transformation of nitrogen and nitrogenous compounds mediated by soil microorganisms is of paramount importance in supplying required forms of nitrogen to the plants and various nutritional classes of organisms in the biosphere. In nature, nitrogen exists in three different forms *viz*. gaseous / gas (78 to 80 % in the atmosphere), organic (proteins, amino acids, chitins, nucleic acids, and amino sugars) and inorganic (ammonia and nitrates).

The nitrogen cycle is the sequence of biochemical changes from free atmospheric N_2 to complex organic compounds in plant and animal tissues and further to simple inorganic compounds (ammonia, nitrate) and eventual release of molecular nitrogen (N_2) back to the atmosphere is called "nitrogen cycle".

In this cycle a part of atmospheric nitrogen (N_2) is converted into ammonia and then to amino acids (by soil microorganisms and plant-microbe associations) which are used for the biosynthesis of a complex nitrogen-containing organic compound such as proteins, nucleic acids, amino sugars, etc. The proteins are then degraded to simpler organic compounds *viz*. peptones and peptides into amino acids which are further degraded to inorganic nitrogen compounds like ammonia, nitrites, and nitrates. The nitrate form of nitrogen is mostly used by plants or may be lost through leaching or reduced to gaseous nitrogen and subsequently goes into the atmosphere, thus completing the nitrogen cycle. Thus, the process of mineralization (conversion of an organic form of nutrients to its mineral /inorganic form) and immobilization (the process of conversion of mineral / inorganic form of nutrient elements into the organic form) are continuously and simultaneously going on in the soil.

Nitrogen cycle

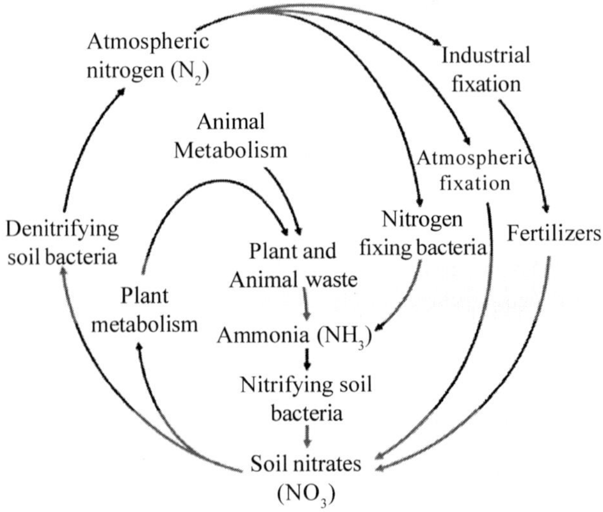

Nitrogen undergoes several transformations that involve organic, inorganic, and volatile forms of nitrogen. A small part of the large reservoir of N_2 in the atmosphere is converted to organic compounds by certain free-living microorganisms or by plant-microbe association that makes the element available to plant growth. The nitrogen present in the proteins or nucleic acids of plant tissue is used by animals. In the animal body, the N is converted to other simple and complex compounds. Upon the death, plants and animals undergo microbial decay and organic N is released as ammonium, which is then utilized by vegetation or is oxidized to nitrate by microorganisms. The nitrate form of N is mostly used by the plants or maybe lost by bacteria reduced to gaseous N_2, which escapes to the atmosphere, thereby completing the cycle.

The Nitrogen cycle mainly includes the following transformations such as

- Mineralization
- Nitrification
- Denitrification
- Immobilization
- Nitrogen Fixation

1. Mineralization

It is the process in which proteins, nucleic acids, amino sugars, and other organic compounds of plant and animal residues are degraded by microorganisms with the eventual liberation of ammonia. This is also called ammonification. A part of the liberated ammonia is assimilated by the microorganisms themselves.

Ammonification occurs in two processes.

a. Proteolysis

b. Deamination

A. Proteolysis

Proteolysis is the process of enzymatic breakdown/hydrolysis of proteins,nucleic acids, and other organic nitrogenous compounds into amino acids by the microorganisms with the help of proteolytic enzymes. The breakdown of proteins is completed in two stages:

1. Proteins are converted into peptides or polypeptides by enzyme proteinases
2. Polypeptides / Peptides are further broken down into amino acids by the enzyme peptidases.

Protein degradation

$$\text{Proteins} \xrightarrow{\text{Proteinases}} \text{Peptides} \xrightarrow{\text{Petidases}} \text{Aminoacids}$$

The most active microorganisms responsible for elaborating the proteolytic enzymes (Proteinases and Peptidases) are *Pseudomonas, Bacillus, Proteus, Clostridium histolyticum, Micrococcus, Alternaria, Penicillium,*etc.

B. Deamination

After proteolysis, the amino compounds are deaminated to yield ammonia with the activity of deaminase enzyme.

Amino acid demination

$$\underset{(Alanine)}{CH_3\text{-}CHNH_2\text{-}COOH} + {}^1/_2\,O_2 \xrightarrow{\text{Deaminase}} \underset{(Pyruvic\ acid)}{CH_3COCOOH} + NH_3$$

Almost all bacteria, actinobacteria and fungi can bring about proteolysis, and the amino acids produced are utilized for the growth of these organisms.

Putrefaction

Ammonification usually occurs under aerobic conditions, while under anaerobic conditions protein decomposition leads to conversion of ammonia into amines and related compounds (eg.) *Clostridium*. The anaerobic decomposition of protein is called putrefaction. These amines are subsequently oxidized in the presence of O_2 to release ammonia.

2. Nitrification

The biological oxidation of ammonium salts (in soil) to nitrites and the subsequent oxidation of nitrites to nitrates is called nitrification. i.e. the biological conversion of N in soil from a reduced to a more oxidized state, called nitrification. This process occurs primarily by a group of aerobic chemoautotrophs called the nitrifiers. While some low levels of heterotrophic nitrification do occur, but rates are low and quantities of nitrate produced are relatively small compared to the chemoautotrophs. All nitrifiers are aerobic, and nitrification proceeds at C: N ratios of less than 20, when N is abundant. In some settings, such as forest litter layers, most nitrification occurs by saprophytic fungi rather than chemoautotrophic bacteria. Nitrification is a two-step process.

First Step: Ammonium Oxidation

Ammonia is oxidized to nitrite.

$$2NH_3 + 1½H_2O_2^- \rightarrow NO_2^- + 2H{+}H_2O \text{ – Nitrosofication}$$

The microorganisms involved are called ammonia oxidizers. This change is brought about by chemoautotrophic bacteria of the genera *Nitrosomonas*, *Nitrosolobus, Nitrosococcus, Nitrosospira*. These bacteria obtain their energy requirement by the oxidation of NH_4^+ to NO^-_2. Among the nitrifiers, *Nitrosomonas* are most important in soils. *Nitrosospira* is an aquatic nitrifier.

Second Step: Nitrite Oxidation

Nitrite is further oxidized to nitrate.

$$HNO_2 + ½\,O_2 \rightarrow HNO_3.$$

The Organisms involved are *Nitrobacter*, *Aspergillus*, *Penicillium*, *Cephalosporium*.

Nitrification inhibitors

The commercial compounds such as pyridines, pyrimidines, amino triazoles, and sulfur compounds like ammonium thiosulfate will inhibit the activity of nitrifying bacteria. Neem oil slows down the nitrification process

Nitrogen is lost from ecosystems mainly after its conversion to $NO_3$and before plant uptake and hence keeping N in the NH_4 form, keeps it from being lost by nitrate leaching and denitrification which are the two principal pathways of unintentional N loss in most ecosystems.

3. Denitrification

The conversion of nitrate and nitrite into molecular N_2 or nitrous oxide through microbial processes is known as denitrification. Certain bacteria are capable of using nitrate as the terminal electron acceptor under anaerobic conditions. This is called **nitrate respiration**. As a consequence of nitrate respiration, NO_3 is reduced to N_2 gas or nitrous oxide. Denitrification leads to the loss of N from the soil. It depletes N, and therefore it is not a desirable reaction. The escape of molecular N into the atmosphere is also known as **volatilization**.

Denitrification occurs mostly in waterlogged anaerobic soils with high organic matter content. Denitrification of bound nitrogen to gaseous N is mediated by numerous species of bacteria, which normally use O_2 as a hydrogen acceptor (aerobically) and, also use nitrates and nitrites (anaerobically). Anaerobic conversion of nitrate into molecular nitrogen is known as nitrate respiration. Bacterial genera that bring about denitrification are *Pseudomonas, Achromobacter, Bacillus, Micrococcus.*

$$2NO^{-}_{3} + 10\ H^{-} \rightarrow N_2 + 4H_2O + 2OH^{-} \quad \text{(or)}$$

$$2NO^{-}_{2} + 6\ H^{-} \rightarrow N_2 + 2H_2O + 2OH^{-} \quad \text{(or)}$$

$$N_2O + 2H^{-} \rightarrow N_2 + H_2O$$

Since nitrates are used as a source of electron acceptor, there is a net loss of N from the soil. This process is termed also as **dissimilatory nitrate reduction**. Many soil bacteria like *Thiobacillus denitrificans* oxidize S (chemoautotrophically) and also reduce nitrate to nitrogen.

$$5S + 6\ KNO_3 + 2\ H_2O \rightarrow 3N_2 + K_2SO_4 + 4KHSO_4 \quad \text{(or)}$$

$$5\ K_2S_2O_3 + 8\ KNO_3 + H_2O \rightarrow 4N_2 + 9\ K_2SO_4 + H_2SO_{4\text{-}}$$

General pathway of denitrification

Nitrate is first reduced to nitrite, which is then transformed into nitrous oxide (NO). The nitrous oxide is converted to N_2 with N_2O as an intermediate.

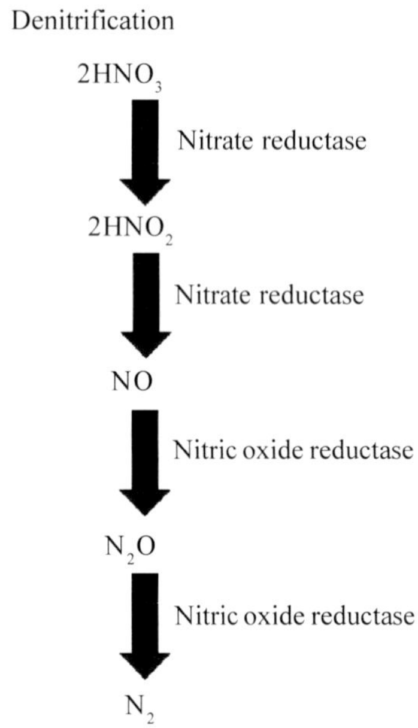

Fallow soils flooded with water are more congenial for denitrification than well-drained and continuously cropped soils. Though it is an undesirable reaction from the point of view of plant nutrition, it has ecological importance. Because without denitrification the supply of N on the earth would have got depleted and NO_3 would have accumulated. High concentration of NO_3 are toxic, denitrification is a mechanism by which some of the N is released back to the atmosphere. It is essential to distinguish nitrification, denitrification as separate processes, and the difference between these are furnished below in the table.

	Nitrification	Denitrification
Aerobic status	Obligate Aerobes	Facultative
Respiratory Classification	Aerobic Respiration	Anaerobic Respiration
Respiratory electron acceptor	O_2	NO_3^-, NO_2^-, others
C-source	Autotroph	Heterotroph
Energy source	Chemolithotroph	Chemoorganotroph
Taxonomic group	Nitrosomonas, Nitrobacter	Variable group; many different genera
Reversibility	Reversible	Not reversible

The most important denitrifying bacteria are *Thiobacillus denitrificans, Micrococcus denitrificans,* and species of *Pseudomonas, Bacillus, Achromobacter, Serrtatia paracoccus,* etc.

4. Immobilization

In nitrogen immobilization, ammonia and nitrate are taken up by microbes and are largely immobilized or made unavailable to plants, depending on the C: N ratios. When N is abundant, both microbes and plants assimilate ammonia and nitrate. In contrast to mineralization, microbial immobilization leads to the biosynthesis of the complex molecules of microbial protoplasm from ammonium and nitrate. Immobilization results in a marked depression of nitrogen uptake by the plant. The mineralization of organic N and the microbial assimilation of inorganic ions proceed simultaneously in soil.

Relationship between the C: N ratios and net immobilization and nitrification are depicted in the table below.

< 20	20-30	> 30
Net gain of NH_4^+ and NO_3^-	Neither gain nor loss	Net uptake of NH_4^+ and NO_3^-

The C: N ratio is used as an indicator to know the step in the nitrogen cycle. The ratios less than 20 means, excess N is present and nitrification proceeds (with a net gain of N). With ratios between 20 and 30, nitrification and immobilization rates are in equilibrium and there is no net gain or loss of N. With a ratio greater than 30, N is limited and net immobilization occurs with uptake (or loss) of N from the active N cycle.

5. Nitrate reduction

Several heterotrophic bacteria *(E. coli, Azospirillum)* are capable of converting nitrates to nitrites and nitrites to ammonia. The complete reversal of nitrification is known as nitrate reduction. Nitrate reduction normally occurs under anaerobic soil conditions (waterlogged soils) and the overall process is as follows:

Nitrate reduction

$$\underset{\text{Nitrate}}{HNO_3 + 4H_2} \xrightarrow{\text{Nitrate reductase}} \underset{\text{Ammonium}}{NH_4} + 3H_2O$$

Nitrate reduction leading to the production of ammonia is called "assimilatory nitrate reduction "since some of the microorganisms assimilate ammonium, through which they synthesis of proteins and amino acid this process is called as assimilatory nitrate reduction.

11

Biological Nitrogen Fixation

Fixation of elemental nitrogen in the atmosphere by the microorganisms through a reductive process into ammonia is called **biological nitrogen fixation** (BNF). Molecular nitrogen (N_2) is the major component (approximately 80%) of the earth's atmosphere. The element nitrogen is an essential part of many of the chemical compounds, such as proteins and nucleic acids, which are the basis of all life forms. However, N_2 cannot be used directly by biological systems to build the chemicals required for growth and reproduction. Before its incorporation into a living system, N_2 must first be combined with the element hydrogen. This process of reduction of N_2 commonly referred to as **nitrogen fixation** (N-fixation).

N_2 is transformed into reactive N via lightning discharge (5 teragrams (Tg) N / year; 1 Tg equals 1012 g) and by biological N fixation (100 -140 Tg N /year). Conversion of dinitrogen to ammonia in the industrial process requires temperatures of 400 - 500° C and high pressures (Haber-Bosch process). Biological nitrogen fixation is much more important as it contributes to more than 70 % of the input in to the world's soil and water nitrogen. It has been estimated that 100 to 175 million tones of nitrogen are fixed annually by the process of biological nitrogen fixation.

The process of N_2 fixation is mediated by the enzyme, called nitrogenase (which reduces N_2 to ammonia). The overall reaction is as follows.

$$N_2 + 3H_2 \xrightarrow{\text{ATP}} 2\,NH_3$$

A variety of prokaryotic organisms have the ability to reduce the atmospheric nitrogen. Bacteria that use N_2 as the sole source of N are called diazotrophs. Some live in a symbiotic relationship with plants of the legume family (*Rhizobium*). Some establish symbiotic relationships with plants other than legumes (*Frankia*). Some exhibit symbiotic relationship with water fern, Cycads and angiosperms (*Nostoc*).Some nitrogen-fixing bacteria live freely in the soil (*Azotobacter*). Free living Nitrogen fixers are both aerobic and anaerobic. Some of them are phototrophs (Blue green algae) and some are heterotrophs .The root nodule associations were the first to be recognized for their ability to fix atmosphere N_2. Rhizobia are the first group of organism realized for its potential of nitrogen fixation.

Nitrogen fixing bacteria are classified according to their mode of fixation.

1. Free living N fixers
2. Associative N fixers
3. Endophytic N fixers
4. Symbiotic N fixers

1a. Free living Aerobic nitrogen fixers

Azotobacter

Beijerinckia

Klebsiella

Cyanobacteria (Blue green algae)

1b. Free living Anaerobic nitrogen fixers

Clostridium

Desulfovibrio

Purple sulphur bacteria

Green sulphur bacteria

2. Associative symbiotic N fixers

Azospirillum

Herbaspirillum

3. Endophytic N fixers

Gluconacetobacter diazotrophicus

Azoarcus

Serratia

Derxia

4. Symbiotic N fixers

Rhizobium (*Rhizobium* – legume association)

Bradyrhizobium (*Bradyrhizobium* – soybean association)

Azorhizobium (*Azorhizobium- Sesbania rostrata* association)

Photorhizobium - Nodulants *Aeschynomene* sp.

Sinorhizobium - fast growing soybean nodulator

Anabaena azollae (*Azolla – Anabaena* association)

Frankia (*Frankia* – Non legume -*Casuarina* association)

Lichens (Fungi – Algae association)

1. Free Living Nitrogen Fixation

The free-living (nonsymbiotic) bacteria, includes the Cyanobacteria (*Anabaena,Nostoc*)and genera such as *Azotobacter*, *Beijerinckia*, and *Clostridium*. In 1960, J. E. Carnahan and his associates first reported the reduction of gaseous nitrogen to ammonia by the cell-free preparations from the bacterium, *Clostridium pasteurianum*. They fix nitrogen without depending the host plants or other organisms.Some of the photosynthetic bacteria also exhibit nitrogen fixation.*Azotobacter* and cyanobacteria are the major contributors for free living N fixation.

a. Azotobacter

These are free-living bacteria, which grow well on a nitrogen free medium. These bacteria utilize atmospheric nitrogen gas for their cell protein synthesis. This cell protein is then mineralized in soil after the death of Azotobacter cells thereby contributing towards the nitrogen availability of the crop plants. Azotobacter is Gram-negative bacteria, polymorphic i.e. they are of different sizes and shapes. Their size ranges from 2-10x1-2.5 µm.Young cell possess peritrichous flegella and are used as locomotive organs. Azotobacter spp., are sensitive to acidic pH, high salts, and temperature above 35° C. Some of the important species of *Azotobacterviz. A. chroococcum, A. vinelandii, A. beijerinckii, A. paspali*, *A. agilis, A. insignis, A. nigricans, A. salinestris, A. macrocytogens* are commonly found in our soils. It has several types of nitrogenases. The basic one is molybdenum-iron nitrogenase. An alternative type contains Vanadium; it is independent of molybdenum ions and is more active than the Mo-Fe nitrogenase at low temperatures. So it can fix nitrogen at temperatures as low as 5°C.

b. Cyanobacteria

Cyanobacteria are oxygenic photosynthetic bacteria, widespread in marine, freshwater and terrestrial environments, and many of them are capable of fixing atmospheric nitrogen. It fixes the ubiquitously available CO_2 and N_2, and therefore cover the demand of the two most important elements.The fixation of N_2 comes at a high metabolic energy cost, but cyanobacteria are phototrophic organisms that use sunlight to cover their energy demand. They follow the strategy is known as spatial separation of oxygenic photosynthesis and nitrogen fixation. Cyanobacteria separate the incompatible processes of oxygenic photosynthesis and N_2 fixation spatially (in different cells) or temporally (during the night), or a combination of both.

N_2 fixation in the ocean is carried out by free living non heterocystous filamentous and unicellular cyanobacteria and by symbiotic cyanobacteria living with microalgae.Heterocystous cyanobacteria are found in freshwater, brackish water, terrestrial environments and symbiotic in plants and algae.These microorganisms display a relatively wide variety of morphologies, such as unicellular, non-heterocytous and heterocytous filamentous strains with the latter showing different types of cells (heterocytes and akinetes)

Forms of Cyanobacteria

Single celled :*Gleotheca*, *Gleocapsa*

Fliamentous non heterocyctous :*Oscillatoria,Tolyphothrix*

Filamentous heterocystous :*Nostoc*, *Anabaena*

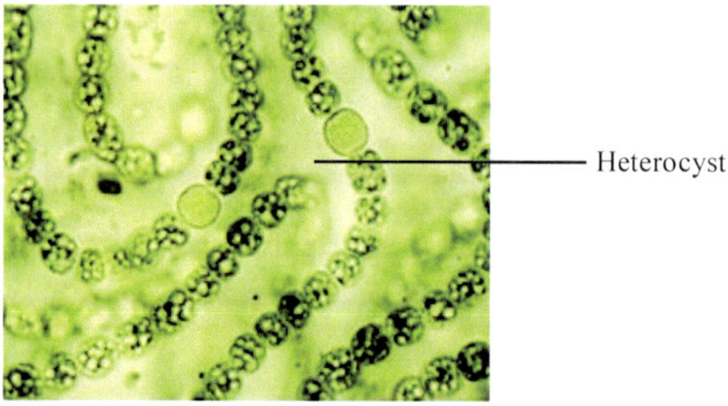

Heterocystous strains are able to form differentiated cells, specialized in nitrogen (N_2) fixation, the heterocysts, and spore-like resting cells, the akinetes. Unicellular and non-heterocystous strains are also able to perform N_2 fixation under certain conditions, despite the absence of specialized cells.Cyanobacteria have a circadian clock that synchronizes metabolic processes and allow for the fixation of N_2.It possess the Mo-Fe cofactor in the nitrogenase.

2. Associative Symbiotic N Fixers

Azospirillum is considered to be an associative symbiotic bacterium because it lives in close association with the root system of plants. Almost all crop plants harbour this bacterium in their roots. These bacteria colonizing the roots not only remain on the root surface but also a sizable population penetrates into the root tissues and live in harmony with the plant. However they do not produce any visible nodules or outgrowth on the roots. In the absence of any plant, *Azospirillum* lives in the soil just like any other microorganism saprophytically. However, when a suitable crop is raised they are attracted towards the root system. Making use of the nutrients exuded by the plants through their root system (root exudates) Azospirillum colonies in the root region in large numbers and fixes substantial amounts of nitrogen. Some of the commonly found species are *A. lipoferum, A. brasilense, A. amazonense, A. hlaopraeferens, A. irakense, A. largimobilis*.

3. Endophytic N fixers

Endophytic bacteria live in plant tissues without causing substantive harm to the host or gaining any benefit other than a non-competitive environment inside the host. They ubiquitously inhabit most plant species and have been isolated from several types of tissues, seeds, roots, stems and leaves. Some endophytes may enhance host growth, nutrient acquisition and may improve the plant's ability to tolerate abiotic stresses, such as drought, and enhance resistance to insects, plant pathogens. Moreover, endophytic bacteria are less vulnerable to competition with other soil microbes for scarce resources and remain protected to various abiotic and biotic stresses.

Endophytic bacteria are better than their rhizospheric and rhizoplanic counterparts in terms of benefiting their host through nitrogen fixation as they can provide fixed nitrogen directly to their host (being sheltered inside the host plants). The partial pressure of oxygen inside the plant tissue is more amenable for efficient nitrogen fixation. Endophytes can contribute upto 47% of nitrogen derived from air.

In addition to *Gluconacetobacter diazotrophicus*, *Herbaspirillum seropedicae*, *H. rubrisubalbicans*, *Azoarcus*, *Serratia*, *Burkholderia*, *Derxia* have also been reported as other endophytic bacteria.

4. Symbiotic N Fixers

Symbiotic nitrogen fixationis part of a mutualistic relationship in which plants provide a niche and fixed carbon to bacteria in exchange for fixed nitrogen. The nitrogen fixation occurs when the microorganisms are in mutual symbiosis with plant hosts .The best-studied example is the association between **legumes** and **bacteria** in the genus *Rhizobium*. A symbiotic relationship in which both partners benefits is called mutualism. A mutualistic symbiosis is an association between two organisms from which each derives benefit. It is usually a long term relationship, and often involves a special structure to house the microbial partner. Each N_2 fixing symbiotic association involves an N_2 fixing prokaryotic organism, the microsymbiont (eg. *Rhizobium*, *Nostoc* or *Frankia*) and a eukaryotic, usually photosynthetic, host (e.g., leguminous or nonleguminous plant, water fern or liverwort).

a. *Rhizobium* - Legume symbiosis

Legumes are the most widely recognized N_2 fixing symbioses because of their importance as a food source. In 1886, Hellriegel and Wilfarth demonstrated that the ability of legumes to convert N_2 from the atmosphere into compounds which could be used by the plant was due to the presence of swellings or nodules on the legume root, and to the presence of particular bacteria within these nodules. Quickly thereafter, Beijerinck in 1888 isolated the bacteria (rhizobia) from legume root nodules and showed they were able to reinfect the legume, form nodules, and fix N_2 in the symbiosis. This symbiosis accounts for around 60% of total biological nitrogen fixation (BNF) inputs in the world agriculture per annum. Legumes have the ability to establish symbiosis with soil bacteria that belong mainly to the genera *Rhizobium*, *Mesorhizobium*, *Bradyrhizobium* and *Azorhizobium*, commonly known as Rhizobia.

Azorhizobium forms stem nodules in *Sesbania*. Stem nodulation is reported in three genera *viz*., *Aeschyonomene*, *Neptunia* and *Sesbania rostrata*.

Mesorhizobium *loti*, formerly known as *Rhizobium loti*, is a Gram negative species of bacteria found in the root nodules of many plant species.

Bradyrhizobium *japonicum* which could nodulate Soybean.

Photorhizobium which could nodulate *Aeschynomene* sp.

Sinorhizobium are also called Ensifer,a fast growing rhizobia.

Cross inoculation groups of *Rhizobium* (CIG)

The success of these legume crops depends upon the efficient functioning of legume plant -*Rhizobium symbiosis.*Cross inoculation groups(CIG) refers the groups of leguminous plants that will develop effective nodules when inoculated with the rhizobia obtained from the nodules from any member of that legume group (Host specificity).

1.	*Rhizobium leguminosarum*	CIG	Host it can nodulate
	bv. viceae	Pea	Peas, Lentils, Vicia
	bv. phaseoli	Bean	*Phaseolus* spp
	bv. trifoli	Clover	*Trifolium* spp
2.	*R. meliloti*	Alfalfa	Alfalfa, Clover, Fenugreek
3.	*R. lupini*	Lupini	*Trifoli*, Lupine,
4.	*R. fredii*	Soybean	Soybean
5.	*R.* spp	Cowpea group	*Vigna*, *Arachis*, *Cajanus*, *Dolichus*, *Sesbania*, *Acacia*, *Prosopis*,
6.	*R.* sp	Chickpea group	Chickpea

b. *Frankia* - Non legume symbiosis / Actinorhizal symbiosis

The term "actinorhiza" refers both to the filamentous bacteria *Frankia*, an actinomycete, and to the root location of the nitrogen fixing nodules. *Frankia* is a genus of Nitrogen fixing filamentous bacteria that live in symbiosis with actionorhizal plants found in the root nodules of non legumes.This genus was originally identified by Brunchorst and considered as a filamentous fungus and named it as *Frankia*in the year 1886.Becking redefined the genus in 1970 as Prokaryotic Actinomycetes and created the family Frankiaceae within the Actinomycetales. *Frankia alni*, *Frankia elagni* are the few named species in this genus, but a great many strains are specific to different plant species. *Frankia* are the gram positive bacteria,which fix nitrogen under both free living and symbiotic conditions.*Frankia* provides the plant with a source of nitrogen, and in return the plant provides the *Frankia* with a carbon source.

Actinorhizal plants represent a diverse group of about 220 species belonging to eight plant families distributed in the three orders

- Fagales (Betulaceae, Casuarinaceae and Myricaceae),
- Rosales (Rosaceae, Eleagnaceae and Rhamnaceae)
- Cucurbitales (Datiscaceae and Coriariaceae)

In arid soils, the species *Casuarina equisetifolia* fixes an average of 15 kg N /ha/ year. But in temperate climates, nitrogen-fixation activity in actinorhizal plants could be similar to the rate of 300 kg /year measured in legumes.

c. Cyanobacterial associations

Cyanobacteria associate with nonvascular plants (mosses, liverworts, hornworts), ferns, cycads, and angiosperms. Cyanobacteria involved in N-fixing symbioses with plants commonly belong to the order Nostocales. They are having Specialized N-fixing cells (heterocysts) and the resting cells (akinetes). The frequency of heterocysts is more when it is in symbiosis compared to free living forms.

i. *Azolla – Anabaena* symbiosis

Azolla fern

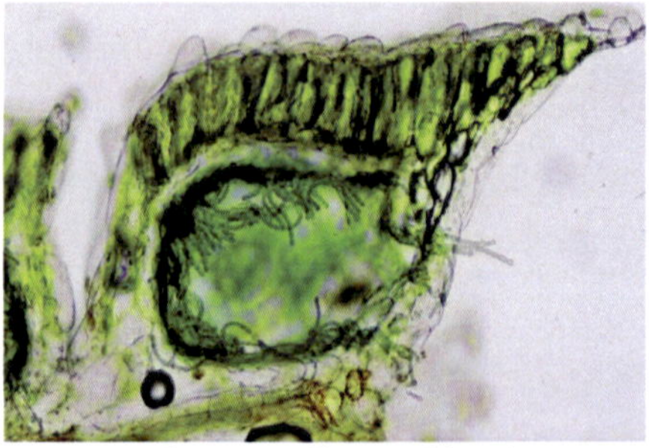

Anabaena azollae symbiont in dorsal lobe of *Azolla* front

The eukaryotic water fern *Azolla* and the prokaryotic endosymbiont*Anabaena azollae* are actively involved in the symbiotic association and fix atmospheric nitrogen. The host plant *Azolla* provides the carbon source sucrose to the cyanobiont while the cyanobiont fix nitrogen and transfers to *Azolla* which in turn multiplies very rapidly. The cyanobiont is associated with dorsal lobes of the leaf from the onset of their development and is never in direct contact with the external environment and *A. azollae* is the only species found in symbiotic association with *Azolla*. The growth and development of the *Azolla* and the *Anabaena* are synchronous.

ii. *Cyanobacteria - Cycad* association

Cycads establish symbioses with filamentous cyanobacteria in highly specialized lateral roots termed "coralloid roots". All of the approximately 150 spp of cycads harbor N-fixing cyanobacteria in coralloid roots.The coralloid roots recorded in all cycad genera are arising from lateral roots and show a marked negative geotropism, and grow laterally and upward toward the surface of the soil. These roots are colonized by cyanobacterial filaments. Symbiotic *Nostoc* filaments show only few modifications in comparison to its free-living counterparts:

a. an increased heterocyst frequency,

b. an increased level of nitrogen fixation, and

c. a transfer of fixed nitrogen from the cyanobacterial to the cycad cells

(a) Coralloid roots

(b) Coralloid roots cross-section

iii. Cyanobacterial - Gunnera Symbiosis

Gunnera are the angiosperms coming under the family Gunneraceae. Cyanobiont (Nostoc) enters the *Gunnera* stem through specialized glands that secrete polysaccharide mucilage.Cyanobiont is held intracellularly (unlike in non-vascular plants,*Azolla,* and cycads), with filaments occupying most of the host cells (filaments surrounded by host cytoplasm). The well known example is *Nostoc punctiformi -Gunnera insignis* association.

Nodule formation in legumes by *Rhizobium*

Rhizobium is the most well known species of a group of bacteria that acts as the primary symbiotic fixer of nitrogen. These bacteria can infect the roots of leguminous plants, leading to the formation of lumps or nodules where the nitrogen fixation takes place. The bacterium supplies a constant source of reduced nitrogen to the host plant and the plant furnishes nutrients and energy for the activities of the bacterium. About 90% of legumes can become nodulated.In soil, the bacteria are free living and motile, feeding on the remains of dead organisms. Free living rhizobia cannot fix nitrogen and they have a different shape from the bacteria found in root nodules. They are regular in structure, appearing as straight rods; in root nodules the nitrogen-fixing form exists as irregular cells called 'bacteroids' which are often club and Y-shaped.

Root nodule formation

Sets of genes in the bacteria control different aspects of the nodulation process. One *Rhizobium* strain can infect certain species of legumes but not others e.g. the pea is the host plant to *Rhizobium leguminosarum* biovar *viciae*, whereas clover acts as host to *R. leguminosarum* biovar *trifolii*, as discussed in CIG of *Rhizobium*. Specificity genes determine which *Rhizobium* strain infects which legume. Even if a strain is able to infect a legume, the nodules formed may not be able to fix nitrogen. Such rhizobia are termed ineffective. Effective strains induce nitrogen-fixing nodules. Effectiveness is governed by a different set of genes in the bacteria from the specificity genes. The initial interaction between the host plant and free-living rhizobia is the release of a variety of chemicals by the root cells into the soil. Some of these encourage the growth of the bacterial population in the area around the roots (the rhizosphere).

Legume effect: The moment when the plant is raised, the rhizobia population build up is very high and this effect is not noticed when non legume is raised. Previously it was thought that homoserine compounds were the triggering this effect. Later it was proved that the signal compounds in the root exudates of legumes are responsible. Signal compounds are not secreted by all plants, only legume alone are producing and they fall into four categories

- Flavones
- Flavanoids
- Isoflavones
- Chalcones

The Signal Compounds are

- Low molecular weight compounds
- Secondary metabolites
- Produced in micromolar quantities
- General structure has benzene ring
- Recognizes Rhizobia
- Eg. Apigenin, Quercetin, Leutinol, Niranginin, Chrysin, etc.

Flavonoids secreted by the root cells activate the nod genes in the bacteria, which then induce nodule formation. The whole nodulation process is regulated by highly complex chemical communications between the plant and the bacteria. Lectins are low molecular carbohydrate binding protein present in almost in all plants and microbes donot produce lectin. Cell wall of root hair contain lectin which helps to attach with exoployscharide of rhizobia and thereby cause lectin binding reaction. Without lectin, cementing could not be possible and it also helps in orientation of rhizobia on root hair. Once binding occurs to the root hair, the bacteria excrete nod factors which stimulate the hair to curl. Rhizobia produces hormones like indole acetic acid (IAA) and makes the root hair curling and deformation, which leads to the loosening of the root hair cells. At the point of infection by rhizobia, root hair cells get hydrolysed which paves way for rhizobium to enter.

*Rhizobium root n*odule development

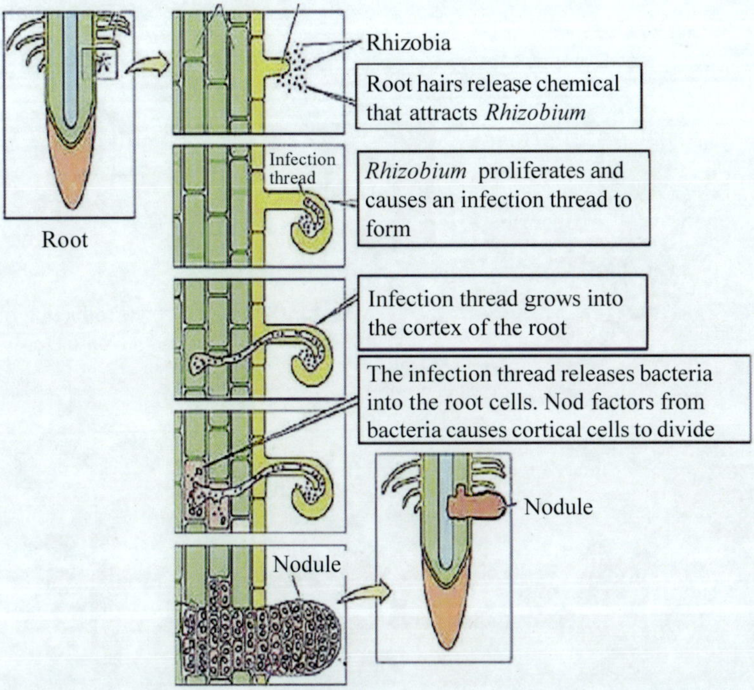

Rhizobia then invade the root through the hair tip where they induce the formation of an infection thread. This thread is constructed by the root cells and not the bacteria and it is formed only in response to infection. The infection thread grows through the root cortex cells and penetrates other root cells nearby often with branching of the thread. The bacteria multiply within the expanding network of tubes, continuing to produce nod factors, which stimulate the root cells to proliferate, eventually forming a root nodule. Within a week of infection small nodules are visible to the naked eye. Each root nodule is packed with thousands of living *Rhizobium* bacteria, most of which are in the misshapen form known as '**bacteroids**'. Portions of plant cell membrane surround the bacteroids,these structures, known as symbiosomes, which may contain several bacteroids or just one, are where the nitrogen fixation takes place.

Mechanism of root nodule development

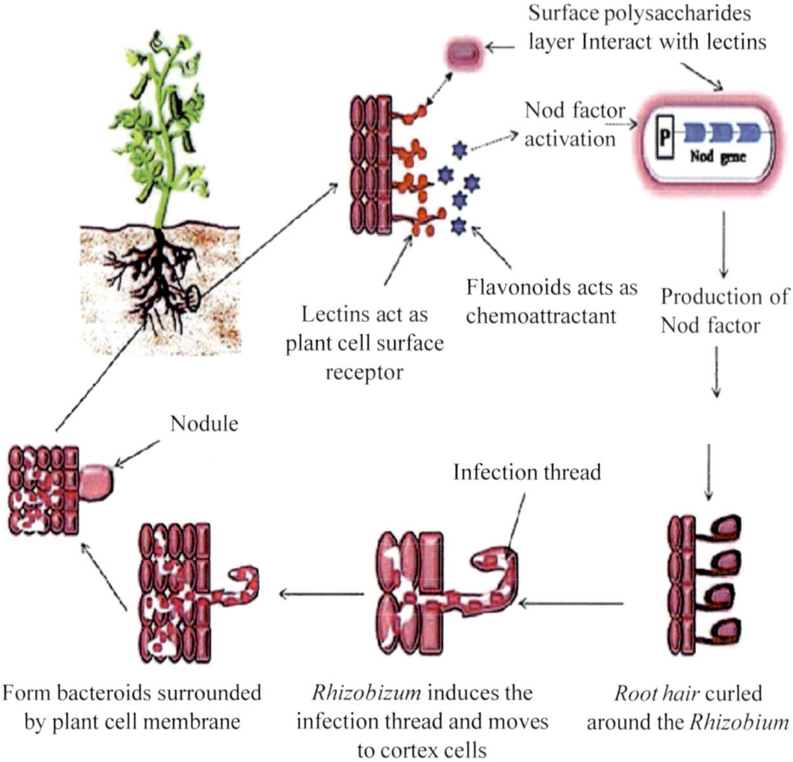

The form in which nitrogen is transported depends on the plant: temperate legumes, which generally form indeterminate nodules, export amides, whereas tropical legumes, which form determinate nodules, export ureides. Actinorhizal plants export mostly amides, with the exceptions of *Alnus*sp and *Casuarina equisetifolia*, which are citrulline exporters. In all cases, ammonium is exported by the microsymbiont as the first product of nitrogen fixation and is assimilated in the cytoplasm of nodule cells via the glutamine synthetase (GS)/glutamate synthase (GOGAT) pathway.

Mechanism of Nitrogen fixation

A relatively small amount of ammonia is produced by lightning. Some ammonia also is produced industrially by the Haber-Bosch process, using an iron-based catalyst, very high pressures and fairly high temperature. But the major conversion of N2 into ammonia, and then into proteins, is achieved by microorganisms in theprocess called nitrogen fixation (or dinitrogen fixation). The total biological nitrogen fixation is estimated to be twice as much as the total nitrogen fixation by non-biological processes.

Biological nitrogen fixation can be represented by the following equation, in which two moles of ammonia are produced from one mole of nitrogen gas, at the expense of 16 moles of ATP and a supply of electrons and protons (hydrogen ions):

$$N_2 + 8H^+ + 8e^- + 16\ ATP \rightarrow 2NH_3 + H_2 + 16ADP + 16\ Pi$$

This reaction is performed exclusively by prokaryotes (the bacteria and related organisms), using an enzyme complex termed nitrogenase. Nitrogenase catalyses the conversion of nitrogen gas to ammonia in nitrogen-fixing organisms. In legumes it only occurs within the bacteroids. The reaction requires hydrogen as well as energy from ATP. The nitrogenase complex is sensitive to oxygen, becoming inactivated when exposed to it. This is not a problem with free living, anaerobic nitrogen- bacteria such as *Clostridium*. Free-living aerobic bacteria have a variety of different mechanisms for protecting the nitrogenase complex, including high rates of metabolism and physical barriers.

Rhizobium controls oxygen levels in the nodule with leghaemoglobin. This red, iron-containing protein has a similar function to that of haemoglobin; binding to oxygen. This provides sufficient oxygen for the metabolic functions of the bacteroids but prevents the accumulation of free oxygen that would destroy the activity of nitrogenase. It is believed that leghaemoglobin is formed through the interaction of the plant and the rhizobia as neither can produce it alone.

Leghaemoglobin

It is synthesized by plant in the nodule tissue when infection thread transform to bacteroids. Leghaemoglobin synthesis switches on when bacteroid formation initiates. It helps to maintain redox potential of nodule, helps in absorption of oxygen and not permitting to destroy nitrogenase, helps to supply nutrients to bacteroids.

The nitrogenase system is present only in prokaryotes. It consists of two dissociating protein components. Component I (also called enzyme I, molybdoferredoxin, and Mo-Fe protein), contains Mo, Fe and S. Component II, (also called enzyme II, azoferrodoxin, azofer and Fe protein), contains Fe and S. Each component by itself has no activity of its own. Combination of the two inactive components reconstitutes nitrogenase activity.

The two components interact as a complex *in vivo*, and one component is required for the stability of the other. The two inactive components have similar physical and chemical properties. Nitrogenase is the diagnostic enzyme for nitrogen-fixing organisms; it consists of two proteins:

(i) **Dinitrogenase:** This component reduces nitrogen (N_2) to ammonia (NH_3). It contains an iron-molybdenum cofactor (abbreviated FeMoCo) that accepts electrons from dinitrogenase reductase.

(ii) **Dinitrogenase reductase**: This component transfers electrons to dinitrogenase. It contains an iron atom that is involved in the redox chemistry. A noteworthy limitation: dinitrogenase reductase is irreversibly inactivated by oxygen. As a result, nitrogen fixation cannot be done in the presence of oxygen.

None of these two components alone can catalyse the reduction of N_2 to NH_3.

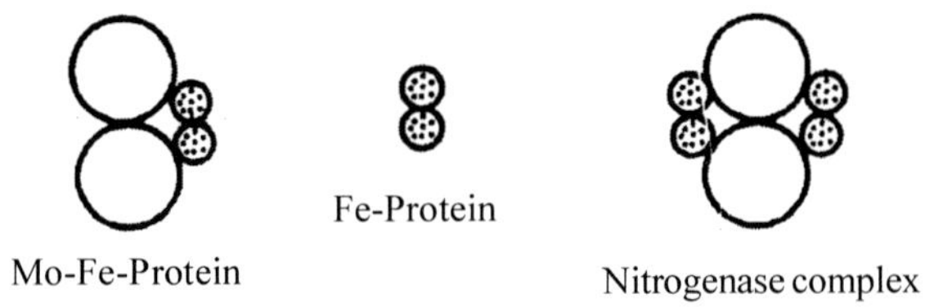

The two subunits are having certain differences in structure as well as function.

Dinitrogenase reductase	Dinitrogenase
Fe, sometimes associated as Fe-S Protein	Subunits-Fe,Mo
Smaller in size	Larger in size
Molecular wt – 64 kD	Molecular wt – 230 kD
Homodimer containing Subunits encoded by gene.	Tetramer containing 2 alpha nifH subunits coded by nifD gene & 2 beta subunits coded by nifK gene
It mediates the electron transfer from donor to acceptor	Function is the convertion of N_2 to NH_3

Nitrogenase is highly sensitive to oxygen. Aerobic organisms face special challenges to nitrogen fixation because nitrogenase is inactivated when oxygen reacts with the iron component of the proteins.

Properties of nitrogenase enzyme

- Highly primitive enzyme
- No substrate specificity for the enzyme
- Oxygen labile
- Looses its activity at Purification
- No inhibition by substrate
- Regulation by its products
- pH optima 6.5 –7.5
- Complex enzyme

There are three different forms of nitrogenase that differ in their requirements for molybdenum (Mo), vanadium (V), or iron (Fe) as a critical metallic component of the cofactor associated with the catalytic site. Most of the nitrogenases contain a Mo cofactor. Nitrogenases with Vanadium/Nickel as cofactor are called unusual nitrogenases.

Energy Requirement for Nitrogen fixation

Biological nitrogen fixation occurs in the presence of the enzyme nitrogenase which is found inside the nitrogen fixing prokaryote. In addition to this enzyme, a source of reducing equivalents (ferredoxin (Fd) or flavodoxin *in vivo*), ATP and protons are required. BNF is the high energy consuming reaction. Mg-ATP is absolutely required as the energy source for this process. The Fe-protein binds with Mg-ATP. It needs two ATPs for the transfer of each electron during the reduction of Nitrogen to Ammonia.ATPs are synthesized during aerobic and anaerobic respirations as well as through phototophosphorylation in different groups of nitrogen fixers.

Aerobic organisms: By the oxidation of respiratory substrates

Anaerobic organisms: Through the 'phosphoroclastic' cleavage of pyruvate to acetate

Photosynthetic organisms: via photophosphorylation

Electron donors

Pyruvic acid and NADPH are the electron donors in most of the cases. Hydrogen gas can also serve as the electron donor. Other sources of electrons are the photolysed water, the assimilated carbon and reduced sulphur compounds. It is also evident that the electron-donating systems could be PS-I, the glycolysis and Krebs cycle and the pentose phosphate pathway.

Electron carriers

Ferridoxin, Flavodoxin, Flavin Mono Nucleotide and Ubiquinone are the electron carriers. Ferridoxins are acidic iron sulphur proteins. Ferridoxins have the lowest redox potentials, 0.42 V, of any biological reductant. It is capable to transfer single electron. Flavodoxins are acidic protein containing the prosthetic group. Flavodoxin is produced instead of ferridoxin when the organism is grown under conditions of limited iron supply.But in some cases both may operate.

Process of Nitrogen fixation

Functioning of nitrogenase enzyme complex

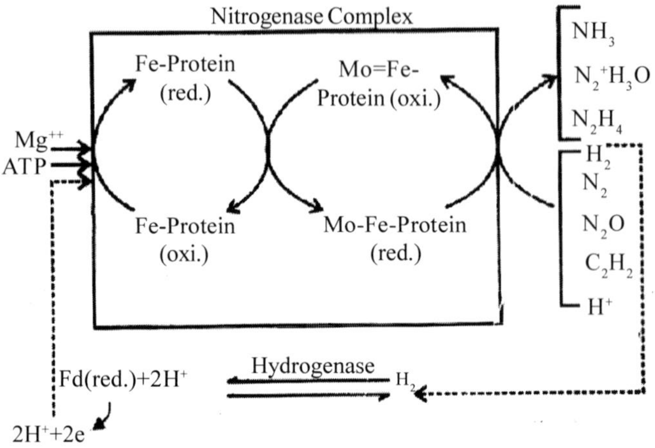

The nitrogenase complex, consisting of di-nitrogenase reductase (Fe-protein) and di-nitrogenase (Mo-Fe-protein) catalyses the nitrogen fixation. The enzyme remains in the cytoplasm of the cell. Electrons are transferred from NADH (formed in the TCA cycle) to the dinitrogenase reductase *via* soluble electron carrier, ferredoxin. Dinitrogenase reductaseis a single electron carrier;it contains two ATP binding sites. It becomes reduced once it receives electron. After the reduction of di-nitrogenase reductase, Mg-ATP molecules bind to it, bringing about a conformational change of the protein, as a result the redox potential of the Fe_4-S_4 cluster is raised from – 0.25 to – 0.40 V.

Then the electrons will be transferred from Fe-S protein to Mo-Fe protein (*i.e.* dinitrogenase). Following the transfer of an electron to Mo-Fe protein,the bound ATP molecules are hydrolysed to ADP and Pi and then released from the enzyme surface (Fe-S protein), restoring the low redox potential of the enzyme which becomes ready again to take up another electron from ferredoxin. Thus, two molecules of ATP are consumed per electron transferred from NADH to di-nitrogenase via di-nitrogenase reductase. Then dinitrogenase transfer electron to N_2 and reduces it to Ammonia via different steps. For each electron transfer two ATP molecules are required. A total of 16ATP molecules are required to provide (transfer) the eight electrons necessary to reduce one N_2 molecule into two NH_3. One mole of H_2 is formed per mole of N_2 transformed to 2 NH_3.

Dinitrogen reduction

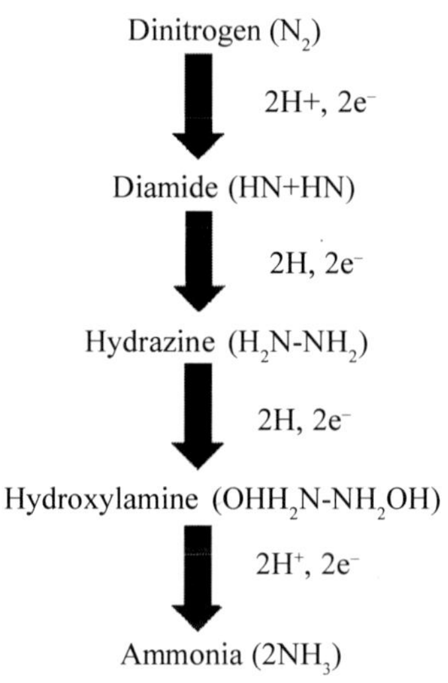

Hydrogen evolution

Nitrogenase forms molecular hydrogen, if no suitable electron acceptor is available. Hydrogen is evolved by many nitrogen fixing bacteria during nitrogen fixation. No hydrogen is, however, evolved when they are grown on NH^{+}_{4}, because nitrogenase synthesis is repressed. Nitrogenase itself can be responsible for H_2 evolution. Nitrogenase receives electrons during nitrogen fixation, and transfers them to the dinitrogen molecule; three electrons (e^-) associate with each atom of nitrogen. Three protons (H^+) are therefore withdrawn from the intracellular medium to neutralize, the negative charge of the electrons. Thus, each dinitrogen molecule yields two molecules of ammonia.

Electron transfer from nitrogenase to molecular nitrogen has a side reaction which is energetically wasteful. In the absence of N_2 and in the presence of ATP and electrons, the electrons combine with protons to yield $H_2$and ATP is hydrolyzed during hydrogen evolution. Even when gaseous nitrogen is available, some electrons are evolved as H_2. The evolved hydrogen is of no apparent use to the plant or the bacteria. Energy derived from metabolism is used in separating free protons (H^+) and electrons (e^-) of $H_2$for use in driving the nitrogen fixation reaction.When the protons and electrons recombine to form H^+, the energy is wasted.

H_2 is a competitive inhibiter of N_2 fixation. Some strains of *Rhizobium* carry the hup gene in Fe Protein in the plasmid. Such strains are able to synthesize hydrogen oxidizing 'uptake hydrogenase'. This enzyme converts molecular hydrogen back to electrons and protons, which arc then reused by nitrogenase. The increased bacterial efficiency would enable the host plant to direct less energy for supporting the bacteria, and more for increasing seed yield. Wild type *Rhizobium* strains capable of oxidizing H_2 through hydrogenase appear to be more efficient in nitrogen fixation than strains without hydrogenase.

Ammonia Assimilation

NH_3, the primary product of N_2 fixation, is toxic to cells in high concentrations. N_2-fixing bacteria possess highly efficient mechanisms of NH_3 assimilation that prevent its accumulation. The following are the three enzymes involved in Ammonia assimilation.

i. Glutamate dehydrogenase (GDH)

ii. Glutamine synthetase (GS)

iii. Glutamate synthase/Glutamine 2 oxoglutarate amino transferase (GOGAT)

In most diazotrophs, the enzyme combination of glutamine synthetase (GS) and glutamate synthase (GOGAT) carries out this task. When ammonia concentration exceeds 0.1 mM NH_4 nitrogen / > 0.5 ppm of NH_3, GDH will be functioning and if it is lower GS-GOGAT operates. In free-living diazotrophs, the nitrogenase-generated ammonium is assimilated into glutamate through the glutamine synthetase/glutamate synthase pathway.Both ATP and reductants are required at this stage of NH_3 assimilation which is an additional energy burden to the N_2-fixing organism. Because BNF is energetically expensive, bacteria do not usually fix N_2 in the presence of reactive N sources. A number of mechanisms have been identified whereby nitrogenase activity is inhibited, and *nif* gene expression is down-regulated in response to increasing levels of NO_3, NH_4, and/or amino acid N in the environment.

$$\text{Glutamate} + NH_4 + \text{ATP} \xrightarrow{\text{GS}} \text{Glutamine} + \text{ADP} + \text{Pi}$$

$$\alpha\text{- ketoglutaric acid} + \text{Glutamine} + NADH_2 \xrightarrow{\text{GOGAT}} \text{Glutamate} + \text{NADPH} + H_2O$$

$$\text{Glutamine} + \text{Glutarate} \xrightarrow{\text{GDH}} 2\ \text{Glutamate}$$

Oxygen protection mechanisms for nitrogenase

Nitrogenase enzyme is sensitive to oxygen.There are certain strategies by which the enzyme is protected from oxygen. There are strategies adopted by the organisms for limiting O_2 access to Nitrogenase.

1. The organisms could avoid O_2 and live in environments which are permanently anaerobic.
2. Physical barrier around nitrogenase prevent O_2 from diffusing to the enzyme.
3. The microorganism by its metabolism, reduce the concentration of O_2 within nitrogenase.
4. They could modify its enzyme in such a manner so as to render its resistant to inactivation by O_2 (conformational protection).
5. The microorganism could simply balance nitrogenase inactivation with the synthesis of new enzyme.

Each and every microorganisms has their own control on nitrogenase activity by which they protect their enzyme. Some of the mechanisms are given below.

i. *Azotobacter* is having respiratory and confirmatory protection against oxygen toxicity. O_2 consumption of *Azotobacter* sp. is 1200 U/ min. Due to this unusually high rate of respiration, it consumes oxygen so fast that the intracellular concentration of oxygen is kept low enough,so that nitrogenase is not inhibited. The moment when O_2 enters the cell, it is sent out as CO_2.It makes the conformational changes in structure of nitrogenase, by which the nitrogen binding sites are buried within the molecule, so that it will not get affected by oxygen supply. Cells are surrounded by copious amounts of exopolysaccharide produced by the cell which protects the nitrogenase from oxygen (Gum protection). Gum protection is also observed in *Beijerinckia* and *Derxia*.

ii. In cyanobacteria heterocyst offer protection against oxygen. Filamentous cyanobacteria have specialized cells called heterocyst that fix nitrogen.Thick walls of heterocyst protect the diffusion of oxygen into these cells. The spatial separation of N_2 fixing site from photosynthetic sites (vegetative cell) eg. *Anabaena*. Also in non heterocystous forms,circadian clock operates, which it fixes nitrogen in dark and carries out photosynthesis during day time.

iii. In *Rhizobium*, the structure of nodules gives protection to nitrogenase.Root nodule tissues and leghaemoglobin protects nitrogenase. An anaerobic environment is maintained in the nodule via the collection of free oxygen by a haeme protein, leghaemoglobin. A thick layer of cells acts as a barrier to oxygen diffusion. Beyond this the oxygen does not diffuse into the nodules easily.

iv. In *Frankia* carotenoids, anthocyanin and tannins give protection form oxygen damage. Production of vesicles also offer protection.

v. Being a facultative anaerobe, *Klebsiella* fixes nitrogen only when it is in an oxygen-free environment.

vi. *Azospirillum* fixesatmospheric nitrogen only in microaerophilic condition.

vii. Some O_2 insensitive nitrogenase were isolated from *Streptomyces thermoautotrophicus*.

Genetics of nodulation and nitrogen fixation

Nod genes

- Genes directing specific steps in nodulation of a legume by a strain of rhizobium are called Nod genes.
- Nod genes are born on large plasmids, called Sym plasmids
- Nod genes located between two clusters of genes for N_2 fixation called **Nif genes.**
- Nod gene consists of 8 genes; nod A,B,C,D,E,F,L,M.
- Among them nod D controls the function of all nod genes

Nif gene

- Genes responsible for N_2 fixation in the microsymbiont are called *Nif* genes.
- 22 genes are involved, arranged in 7 /8 clusters. *Nif* Q,B,A,L,F,M, Z,W,V,S,U,X,N,E,Y,T,K,D,H,J are the Nif genes and each gene has the control over the nitrogenase enzyme and nitrogen fixation.
- Nif K D H – control Nitrogenase enzyme complex and considered vital for N fixation.
- Nif A is having control over Nif K D H

Factors limiting biological nitrogen fixation

1. Edaphic Factors
2. Climatic Factors
3. Biotic Factors

Interactions between the microsymbiont and the plant are complicated by edaphic, climatic, and management factors. A legume-*Rhizobium* symbiosis might perform well in a loamy soil but not in a sandy soil. It may perform under tillage but not in no-till plots. These factors affect the microsymbiont, the host-plant, or both.

1. Edaphic Factors

Edaphic factors relate to the soil. The six main edaphic factors limiting biological nitrogen fixation are: excessive soil moisture, drought, soil acidity, P deficiency,excess mineral N, and deficiency of Ca, Mo, Co and B.

a. Excessive moisture

Excessive moisture and waterlogging prevent the development of root hair and sites of nodulation, and interfere with a normal diffusion of O_2 in the root system of plants. *Sesbania rostrata* and *Aeschynomene* sp. can actively fix N_2 under these conditions because they are located on the plant stems, rather than on the roots.

b. Drought

Drought reduces the number of rhizobia in soils, and inhibits nodulation and N_2 fixation. Prolonged drought will promote nodule decay. Deep-rooted legumes exploiting moisture in lower soil layers can continue fixing N_2 when the soil is drying.

c. Soil acidity

Soil acidity and related problems of Ca deficiency and aluminum and manganese toxicity adversely affect nodulation, N_2 fixation and plant growth.

d. Phosphorus deficiency

It is common problem in tropical regions and reduces nodulation, N_2 fixation and plant growth. Identification of plant species adapted to low-P soils is a good strategy to overcome this soil constraint. The role of mycorrhizal fungi in increasing plant P uptake with beneficial effects on N_2 fixation has been reported. The addition of P-solubilizing microorganisms, particularly of the genera *Pseudomonas, Bacillus, Penicillium,* and *Aspergillus* can solubilize rock phosphate and organically bound soil P (which constitutes 95 - 99% of the total phosphate in soils).

e. Mineral nitrogen

It inhibits the *Rhizobium* infection process and also inhibits N_2 fixation. The former problem probably results from impairment of the recognition mechanisms by nitrates, while the latter is probably due to diversion of photosynthates toward assimilation of nitrates. Some strains of *Rhizobium,* and particularly stem-nodulating *Azorhizobium caulinodans,* fix N_2 actively even when plants are growing in high-N soils (e.g., in the presence of 200 kg fertilizer N ha^{-1}) . Application of large quantities of fertilizer N inhibits N_2 fixation, but low doses (<30 kg N ha^{-1}) of fertilizer N can stimulate early growth of legumes and increase their overall N_2 fixation. The amount of this starter N must be defined in relation to available soil N.

f. Various microelements Cu, Mo, Co, B are necessary for N_2 fixation. Some of these are components of nitrogenase for example Mo.

2. Climatic Factors

The two important climatic determinants affecting BNF are temperature and light.

a. Temperature

Extreme temperatures affect N_2 fixation adversely. This is easy to understand because N_2 fixation is an enzymatic process. However, there are differences between symbiotic systems in their ability to tolerate high (>35°C) and low (<25°C) temperatures.

b. Light

The availability of light regulates photosynthesis, upon which biological nitrogen fixation operates. This is demonstrated by diurnal variations in nitrogenase activity. A very few plants can grow and fix N_2 under shade (e.g., *Flemingia congesta* under plantain canopy). In alley farming, if trees are planted with food crops like cassava, their nitrogen fixation and growth will be reduced due to shading. Early growth of legume trees is slow and they cannot compete successfully for light.

3. Biotic Factors

Among biotic factors, the absence of the required rhizobia species constitutes the major constraint in the nitrogen fixation process. The other limiting biotic factors could be:

a. Excessive defoliation of host plant

Defoliation (e.g., pruning and lopping) decreases the photosynthetic ability of legumes. It impairs N_2 fixation and can lead to nodule decay. For perennial legumes, nodule decay sheds a high number of rhizobia in the root zone. When new roots develop in subsequent vegetative cycles, nodulation of the legume is expected to improve.

b. Crop competition

Intercropping legumes with non-leguminous crops can result in competition for water and nutrients. This competition can affect N_2 fixation negatively. However, it has been shown that when mineral N is depleted in the root zone of the legume component by the non-leguminous intercrops, N_2 fixation of legumes may be promoted.

c. Insects and nematodes

Insects and nematodes have also been reported to interfere with nodule formation, development, and functions.

12

Phosphorus Cycle and Transformations

Phosphorus is only second to nitrogen as an inorganic nutrient required by both plants and microorganisms. It stimulates growth of young plants, giving them a good and vigorous start. Phosphorus management and nutrition has both economic and environmental implications. Phosphate constitutes nearly 0.1% of the earth's crust. They occur in soil in inorganic and organic forms. The organic phosphorus containing compounds are derived from plants and microorganisms and are composed of nucleic acids, phospholipids, lecithin, phytin and related compounds. In soil, 15-85 % of the total P is organic and soils rich in organic matter contain abundant organic P. The inorganic forms are derived from parent rocks or through fertilizers application and manuring with bone meal. They are soluble in water when present as phosphates of Na, K, Ca, Mg etc. Phosphorus exists mainly as apatites, with the basic formula $M_{10} (PO_4)_6 X_2$. Commonly the mineral (M) is Ca, less often Al or Fe. The anion (X) is either F^-, Cl^-, OH^- or CO^{2-}_3. Diverse combinations of M and X results in 200 forms of P.

It is a major constituent of nucleic acids in all living systems essential in the accumulation and release of energy during cellular metabolism. This element is added to the soil in the form of chemical fertilizers, or in the form of organic phosphates present in plant and animal residues. In cultivated soils it is present in abundance (i.e. 1100 kg/ha), but most of which is not available to plants, only 15 % of total soil phosphorus is in available form. Both inorganic and organic phosphates exist in soil and occupy a critical position both in plant growth and in the biology of soil.

Insoluble inorganic compounds of phosphorus are unavailable to plants, but many microorganisms can bring the phosphate into solution. Soil phosphates are rendered available either by plant roots or by soil microorganisms through secretion of organic acids (eg. lactic, acetic, formic, fumaric, succinic acids etc). Thus, phosphate-dissolving / solubilizing soil microorganisms (eg. species of *Pseudomonas, Bacillus, Micrococcus, Mycobacterium, Flavobacterium, Penicillium, Aspergillus, Fusarium* etc.) plays important role in correcting phosphorus deficiency of crop plants. They may also release soluble inorganic phosphate (H_2PO_4), into soil through decomposition of phosphate-rich organic compounds.

Solubilization of phosphate by plant roots and soil microorganisms is substantially influenced by various soil factors, such as pH, moisture and aeration.In neutral or alkaline soils solubilization of phosphate is more as compared to acidic soils. Many phosphates solubilizing microorganisms are found in close proximity of root surfaces and may appreciably enhance phosphate assimilation by higher plants.

By their action, fungi bacteria and actinobacteria make available the organically bound phosphorus in soil and organic matter and the process is known as mineralization. On the other hand, certain microorganisms especially bacteria assimilate soluble phosphate and use for cell synthesis and on the death of bacteria, the phosphate is made available to plants. A fraction of phosphate is also lost in soil due to leaching. One of the ways to correct deficiency of phosphorus in plants is to inoculate seed or soil with preparations containing phosphate - solubilizing microorganisms along with phosphatic fertilizers.

Sources of Phosphate

1) Greatest reservoirs of phosphate are the insoluble ferric and calcium phosphates in rocks (rock phosphate) in combination with calcium, iron and aluminium. Slow process of weathering releases phosphates to the soil.

2) Phosphates are also added to the soil in the form of artificial fertilizers.

3) Some quantity of phosphorus is deposited in deep sediments in oceans and is brought to the cycle by upwelling (but not extensive).

Microorganisms are known to bring a number of transformations of phosphorus, these include:

(i) Altering the solubility of inorganic compounds of phosphorus,

(ii) Mineralization of organic phosphate compounds into inorganic phosphates,

(iii) Conversion of inorganic, available anion into cell components i.e. an immobilization process

(iv) Oxidation or reduction of inorganic phosphorus compounds. Of these mineralization and immobilization are the most important reactions / processes in phosphorus cycle.

Phosphorus cycle

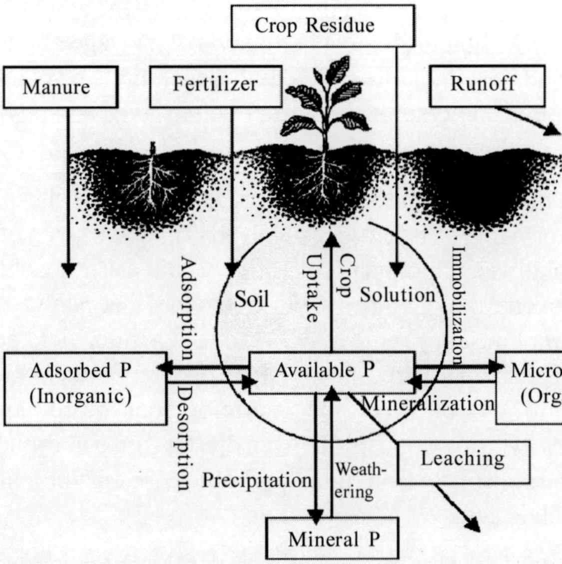

1. Solubilization of inorganic phosphorus

In cultivated soil, P is present abundantly about 1100 kg/ha, but most of them as not available to plants; only about 1% of the total P is in available form. Inorganic phosphorus is negatively charged in most soils. Because of its particular chemistry, phosphorus reacts readily with positively charged iron (Fe), aluminum (Al), and calcium (Ca) ions to form relatively insoluble substances. When this occurs, the phosphorus is considered fixed or tied up. The solubility of the various inorganic phosphorus compounds directly affects the availability of phosphorus for plant growth. The solubility is influenced by the soil pH. Soil phosphorus is most available for plant use at pH values of 6 to 7. When pH is less than 6, plant available phosphorus becomes increasingly tied up in aluminum phosphates. As soils become more acidic (pH below 5), phosphorus is fixed as iron phosphates.

Insoluble inorganic compounds of P are largely unavailable to plants, but many microorganisms can bring the PO_4 into solution. *Pseudomonas*, *Microoccus, Bacillus*, *Fusarium*, *Aspergillus* solubilize calcium salts, iron, aluminum, magnesium phosphate. P is solubilized by the production of organic acids. The acids convert $Ca_3\ (PO_4)_2$ to di and monobasic phosphates and releases P to plants. These organisms secrete high amount of phosphatase enzymes which also solubilize bound form of phosphates.

2. Mineralization of organic phosphorus

Organic form of P is the larger reservoir of P in soil. By the action of bacteria, fungi and actinobacteria, bound element in remains of the vegetation and in soil organic matter is made available to succeeding generations of plants. Among the organic phosphorus compounds, lecithin, nucleic acids and phytin occupy prominent place. Lecithin contains 9.39 % P_2O_5, 1.6% N and 65.36% C.

Mineralization is a process of conversion of organic forms of phosphorus into inorganic available forms of P. The enzyme involved in mineralization of phosphate from organic phosphorus compound is collectively called as Phosphatases.The phosphatase group of enzymes includes phytase enzymes that catalyze the release of phosphate from phytin. They remove PO_4-s, one at a time, yield penta, tetra, di and mono PO_4 and then finally free inositol. Phosphonomoesterases - hydrolyse phosphate from monoester forms of P-nucleotides or phospholipids. Phosphodieasterases hydrolyse phosphate from diester forms of P-nucleic acids. Phytases - hydrolyse phosphate from inositol P. Nuclease enzymes that liberate phosphate from nucleic acids.

These enzymes are produced by 70–80% of the microbial population, including bacteria such as *Bacillus megaterium*, *B. subtilis*, *Serratia* spp., *Pseudomonas* sp., *Proteus* spp., *Arthrobacter* spp., and *Streptomyces* spp. and fungi such as *Aspergillus* spp., *Penicillium* spp., *Rhizopus* spp., Mycorrhizal (fungi) are also able to mineralize the organic forms of P and increases P uptake by the plants. Once P is mineralized, it can be taken up by plants, immobilized by the microbial biomass, precipitated in inorganic complexes, or sorbed to mineral surfaces.

Mineralization of phosphate is generally rapid and more in virgin soils than cultivated land. Mineralization is favored by high temperatures (thermophilic range) and more in acidic to neutral soils with high organic phosphorus content.

3. Immobilization

It is a process of assimilation of P into microbial nucleic acids, phospholipids or other protoplasmic substances is called immobilization. It leads to the accumulation of non-utilizable forms of the element. Soil microorganisms can cause fixation or immobilization of P, either by promoting the formation of inorganic precipitates or by assimilation into organic cell constituents or intracellular polyphosphate granules. P accounts for 0.5-1.0% of fungus mycelium and 1.0 to 3.0% of the dry weight of the bacteria and actinobacteria.

The extent of immobilization of P is affected by the C:P ratio of the organic materials being decomposed and the amount of available P in solution. If insufficient P is available in the substrate for assimilation of the substrate C, inorganic P from the soil solution will be used and net immobilization occurs. Generally, a C:P ratio 200 will result in net mineralization; C:P ratio 300 results in net immobilization, and C:P ratios between 200 and 300 result in little net change in soluble P concentrations.

4. Oxidation reduction reactions

Several soil bacteria and fungi have been shown to be capable of oxidizing reduced phosphorus compounds (e.g., phosphite, hypophosphite) either aerobically or anaerobically.

Oxidative process

Biological oxidation of reduced phosphorus compounds into oxidized state.

Phosphite ($HPO_3^{=}$) is oxidized to phosphate. A number of hetertrophic organisms (bacteria, fungi and actinobacteria) utilize phosphite as sole P source. Hypophosphites ($HPO_2^{=}$) can also be oxidized to phosphate by heterotrophs.

$$HPO_3^{=} \rightarrow HPO_4^{=}$$

$$HPO_2^{=} \rightarrow HPO_4^{=}$$

Reductive process

Reductive pathway has also been functioned. PO_4 is reduced to phosphite and hypophosphite.

$$H_3PO_4 \rightarrow H_3PO_3 \rightarrow H_3PO_2$$

Clostridium.butyricum, *E. coli* form phosphite and hypophosphite from orthophosphate. It is biochemically analogue to the process of denitirification. P exist in an organic form in the protoplasm on the death of living organism, this (P) is changed to inorganic phosphoric acid. This is soon converted into insoluble salts of Ca, Fe, Mg and Al. Phosphorus thus alternates between organic and inorganic, and soluble and insoluble forms.

Phosphate Solubilizing Microorganisms (PSM)

A large number of heterotrophic and autotrophic soil micro organisms known to have the capacity to solubilize inorganic phosphates through their metabolic activities directly (or) indirectly.These organisms were found to mineralize organic phosphorous to soluble and make it available to crop plants. However, PSM vary in their efficiency in solubilizing insoluble phosphates.

Bacteria

Bacillus sp.
Bacillus megnaterium
Bacillus circulans
Bacillus pulvifaciens
Bcillus subtilis
Bacillus mycoides
Bacillus mesentericus
Bacillus flourescens
Pseudomonas sp.
Pseudomonas putida
Pseudomonas liquifaciens
Pseudomonas calcis
Pseudomonas rathonia
Escherichia freundii
Escherichia intermedia
Escherichia intermedia
Xanthomonas sp.
Flavobacterium spp.
Brevibacterium spp.
Serratis spp.
Alcaligenes spp.
Achromobacter spp.
Areobacter aerogenes
Ervinia spp.
Nirosomonas spp.
Thiobacillus thiooxidants

Fungi

Aspergillus spp.
Aspergillus niger
Aspergillus flavus
Aspergillus fumigatus
Aspergillus terreus
Aspergillus awamori
Penicillium spp:
Penicillium lilacinum
Penicillium digitatum
Fusarium spp
Fusarium oxysporum
Currularia lunata
Humicola spp
Sclerotium rolfsii
Pythium spp
Acrothecium spp
Phoma spp
Mortierella sp
Paecilomyces sp
Rhizoctonia sp
Cunninghamella sp

Cyanobacteria

Anabaena sp
Calothrix
Nostoc sp
Scytonema sp
Tolypothrix

Actinobacteria

Actinomyces sp.
Streptomyces

Yeast

Candida sp
Torula
Rhodotorula

Mechanisms of Phosphate Solubilization

1. Production of organic acids

The microbiological means by which insoluble phosphorus compounds are solubilized is by the production of acids. The organic and inorganic acids convert tricalcium phosphates to the di- and monobasic phosphates with the net result of an enhanced availability of the element to the plant. The type of organic acid produced and their amounts differ with different microorganisms.They produce monocarboxylic acids (acetic, formic, lactic, glutamic, glycolic acids) and Dicarboxylic (malic, maleic acid) and Tricarboxylic acids (citric acid). These acids lower the pH, helps in the formation of stable complexes with cation such as ca ++, Mg ++, Fe ++ and Al ++. This prevents the formation of phosphate ions.

Some organisms produced inorganic acids like H_2SO_4 (*Thiobacillus* sp.) and HNO_3 (*Nitrosomonas* and *Nitrobacter*)

2. Production of phosphatase enzymes

These organisms secrete high amount of phosphatase enzymes, which are non specific and hydrolyzing enzymes. Based on pH optima, they divide into Acid phosphatase and Alkali phosphatase. They may be grouped as

1. Phosphomonoesterase
2. Phosphodiestarase
3. Phytases

3. Production of chelating substances

The chelating materials can form the complex with Ca, Fe or Al and release the phosphate from insoluble form. Tricalcium phosphate $Ca_3(PO_4)_2$ is unavailable, when excess Ca is chelated with oxalic acids (Ca oxalate) 2 Ca removed and P will be available.Citric acids destabilize the PO_4 from minerals and ortho phosphate is liberated. α - keto glutaric acid is a powerful chelator of Ca.

4. H_2S production

Due to the reduction of S containing aminoacids many heterotrophic bacteria or sulphate reducing bacteria produces H_2S which reduces ferric phosphate to ferrous sulphids and release the phosphate.

5. Reduction

Ferric forms Fe^{3+} is insoluble and they are reduced by these organisms to Fe^{2+} which is soluble. Under anaerobic conditions they reduce insoluble ferric phosphate to soluble ferrous sulphide with the release of P.

$$H_2S + FePo_4 \rightarrow FeS + H_2Po_4$$

6. CO_2 production

It helps in solubilization of insoluble phosphates by forming carbonates.

7. Siderophore production

Siderophores are low molecular weight iron-loving compounds that hold Fe^{3+} and helps in Fe acquisition. Siderophores bind with iron and prohibit its reaction with soluble phosphates and release phosphate fixed as ferrous phosphate, which is important in acid soils.

13

Sulphur Cycle

Microbial transformations of sulfur (S) are even more complex than those of nitrogen because of the large number of oxidation states of S and the fact that several transformations of S also occur abiotically. Although a number of oxidation states of S are possible, only three are significant in nature, 2 (sulfhydryl, R–SH, and sulfide, HS-), 0 (elemental sulfur, S0), and 6 (sulfate, SO_4 2-). The bulk of S Earth's S is in sediments and rocks in the form of sulfate minerals, primarily gypsum ($CaSO_4$) and sulfide minerals (pyrite, FeS2), but the oceans constitute the most significant reservoir of SO_4^{2-} in the biosphere. A significant amount of S, in particular sulfur dioxide (SO_2), enters the S cycle from human activities, primarily the burning of fossil fuels.

Hydrogen Sulfide and Sulfate Reduction

A major volatile S gas is hydrogen sulfide (H_2S). Hydrogen sulfide is produced from bacterial sulfate reduction or is emitted from sulfide springs and volcanoes. Although H_2S is volatile, different forms exist depending on pH: H_2S predominates below pH 7 and the nonvolatile HS- and S^{2-} predominate above pH 7. Collectively, H_2S, HS, and S_2 are referred to as "sulfide." Sulfate-reducing bacteria are a large and highly diverse group and are widespread in nature. However, in anoxic habitats such as freshwater sediments and many soils, sulfate reduction is SO_4^{2-} limited. Moreover, because organic electron donors (or H_2, which is a product of the fermentation of organic compounds) are needed to support sulfate reduction, it only occurs where significant amounts of organic material are present.

In marine sediments, the rate of SO_4^{2-} reduction is typically C limited and can be greatly increased by an influx of organic matter. This is important because the disposal of sewage or garbage in the oceans or coastal regions can trigger sulfate reduction. Hydrogen sulfide is toxic to many plants and animals and therefore its formation is potentially detrimental (sulfide is toxic because it combines with the iron of cytochromes and blocks respiration). Sulfide is commonly detoxified in nature by combination with iron, forming the insoluble minerals FeS (pyrrhotite) and FeS_2 (pyrite). The black color of sulfidic sediments or sulfate-reducing bacterial cultures is due to these metal sulfide minerals.

Sulphur cycle

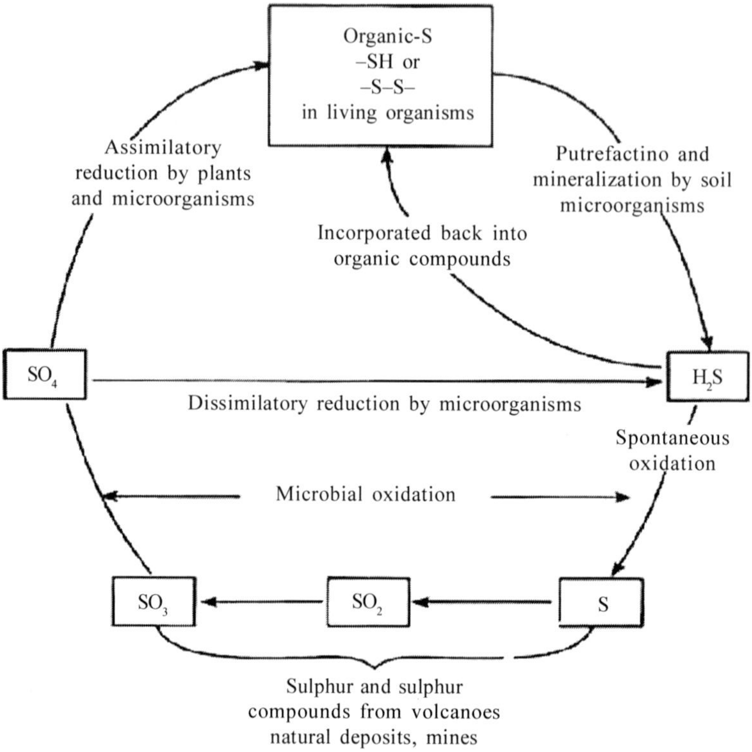

Sulfide and Elemental Sulfur

Oxidation–reduction

Under oxic conditions, sulfide rapidly oxidizes spontaneously at neutral pH. Sulfur-oxidizing chemolithotrophic bacteria, most of which are aerobes, can catalyze the oxidation of sulfide. However, because of the rapid spontaneous reaction, microbial sulfide oxidation is significant only in areas where H_2S emerging from anoxic environments meets air. Where light is available, there can be anoxic oxidation of sulfide, catalyzed by the phototrophic purple and green sulfur bacteria

Elemental sulfur is chemically stable but is readily oxidized by sulfur-oxidizing chemolithotrophic bacteria such as *Thiobacillus* and *Acidithiobacillus.* Because SO is insoluble, the bacteria that oxidize it must attach to the SO crystals to obtain their substrate. The oxidation of SO forms sulfuric acid (H_2SO_4), and thus SO oxidation characteristically lowers the pH in the environment, sometimes drastically. For this reason, SO is sometimes added to alkaline soils as an inexpensive and natural way to lower the pH, relying on the ubiquitous sulfur chemolithotrophs to carry out the acidification process. Elemental sulfur can be reduced as well as oxidized. The reduction of SO to sulfide (a form of anaerobic respiration) is a major ecological process, especially among hyperthermophilic archaea. Although sulfate-reducing bacteria can also reduce SO, in sulfidic habitats most SO is reduced by the physiologically specialized sulfur reducers, organisms that are incapable of SO_4^{2-} reduction. The habitats of the sulfur reducers are generally those of the sulfate reducers, so from an ecological standpoint, the two groups form a metabolic guild unified by their formation of H_2S.

Organic sulfur compounds

In addition to inorganic forms of S, several organic S compounds are also cycled in nature. Many of these foul-smelling compounds are highly volatile and can thus enter the atmosphere.The most abundant organic S compound in nature is dimethyl sulfide (CH_3-S-CH_3); it is produced primarily in marine environments as a degradation product of dimethylsulfoniopropionate, a major osmoregulatory solute in marine algae. This compound can be used as a carbon source and electron donor by microorganisms and is catabolized to dimethyl sulfide and acrylate. The latter, a derivative of the fatty acid propionate, is used to support growth.

Dimethyl sulfide released to the atmosphere undergoes photochemical oxidation to methanesulfonate (CH_3SO_3), SO_2, and SO_4^{2-}. By contrast, CH_3—S—CH_3 produced in anoxic habitats can be microbially transformed in at least three ways: (1) by methanogenesis (yielding CH_4 and H_2S), (2) as an electron donor for photosynthetic CO_2 fixation in phototrophic purple bacteria (yielding dimethyl sulfoxide, DMSO), and (3) as an electron donor in energy metabolism in certain chemoorganotrophs and chemolithotrophs (also yielding DMSO). DMSO can be an electron acceptor for anaerobic respiration, producing CH_3—S—CH_3. Many other organic S compounds affect the global sulfur cycle, including methanethiol (CH_3SH), dimethyl disulfide (H_3C—S—S—CH_3), and carbon disulfide (CS_2), but on a global basis, CH_3—S—CH_3 is the most significant.

Sulphur transformations

Sulphur is the most abundant and widely distributed element in the nature and found both in free as well as combined states. Sulphur, like nitrogen is an essential element for all living systems. In the soil, sulphur is in the organic form (sulphur containing amino acids-cystine, methionine, proteins, polypeptides, biotin, thiamine etc) which is metabolized by soil microorganisms to make it available in an inorganic form (sulphur, sulphates, sulphite, thiosulphale, etc) for plant nutrition.

Of the total sulphur present is soil only 10-15% is in the inorganic form (sulphate) and about 75-90 % is in organic form.

Cycling of sulphur is similar to that of nitrogen. Transformation / cycling of sulphur between organic and elemental states and between oxidized and reduced states is brought about by various microorganisms, specially bacteria. Thus "the conversion of organically bound sulphur to the inorganic state by microorganisms is termed as mineralization of sulphur". The sulphur / sulphate, thus released are either absorbed by the plants or escapes to the atmosphere in the form of oxides.

Various transformations of the sulphur in soil results mainly due to microbial activity, although some chemical transformations are also possible (eg. oxidation of iron sulphide) the major types of transformations involved in the cycling of sulphur are:

1. Mineralization
2. Immobilization
3. Oxidation
4. Reduction

1. Mineralization

The breakdown/decomposition of large organic sulphur compounds to smaller units and their conversion into inorganic compounds (sulphates) by the microorganisms. The rate of sulphur mineralization is about 1.0 to 10.0 percent/year.

2. Immobilization

Microbial conversion of inorganic sulphur compounds to organic sulphur compounds.

3. Oxidation

Oxidation of elemental sulphur and inorganic sulphur compounds (such as H_2S, sulphite and thiosulphale) to sulphate (SO_4) is brought about by chemoautotrophic and photosynthetic bacteria.

When plant and animal proteins are degraded, the sulphur is released from the amino acids and accumulates in the soil which is then oxidized to sulphates in the presence of oxygen and under anaerobic condition (water logged soils) organic sulphur is decomposed to produce hydrogen sulphide (H_2S). H_2S can also accumulate during the reduction of sulphates under anaerobic conditions which can be further oxidized to sulphates under aerobic condition.

a. $2S + 3O_2 + 2H_2O \longrightarrow 2H_2SO_4 \longrightarrow 2H(+) + SO_4$ (Aerobic)

b. $CO_2 + 2H_2S \longrightarrow (CH_2O) + H_2O + 2S$

OR

$H_2 + S + 2\ CO_2 + H_2O \longrightarrow H_2SO_4 + 2(CH_2O)$ (anaerobic)

Sulphur oxidizers are categorized into three groups :

1. Photoautotrophs
2. Chemolithotrophs
3. Heterotrophs

1. Photoautotrophs

This group included the species of green and purple sulfur bacteria. Based on the colour of the carotenoid pigment they are named as purple and green S bacteria. They are reported to oxidize sulphur in aquatic environment. Theses organisms use H_2S as electron donor during photosynthesis in contrast to plants which use H_2O as electron donor and produce oxygen. They are commonly found in mud and stagnant waters containing H_2S and exposed to light. They are also found under extreme conditions of high salinity or temperature and occur in S hot springs and saline lakes as a colored layer under salt deposits.

Purple sulphur bacteria

They belong to Chromatiacea family and generally found in illuminated anoxic zones of lakes and other aquatic habitats where hydrogen sulfide accumulates. They also found in "sulfur springs" - where geochemically or biologically produced hydrogen sulfide can trigger the formation of blooms of purple sulfur bacteria. Anoxic conditions are required for photosynthesis. They can use S, H_2S, Thiosulphides as electron donors.

Examples: *Chromatium*, *Rhodopseudomonas*, *Rhodomicrobium*

Green sulfur bacteria

They belong to cholorobiaceae family. They use sulphide ions, hydrogen or ferrous iron as an electron donor. Elemental sulphur deposited outside the cell may be further oxidized. Examples: *Chlorobium*, *Chloropseudomonas*, *Chlorobacterium*, *Ancalochloris*, *Chlorobaculum*

2. Chemolithotrophs

They are the members of Thiobacilli. *Thiobacillus ferrooxidans* and *T. thiooxidans* are important organisms involved in oxidation of elemental sulphur to sulphates. *Thiobacillus denitrificans* can carry out the S oxidation with NO_3 reduction.

Examples: *Thiobacillus*, *Thiomicrospira,Beggiatoa*, *Thiothrix* and *Thioloca*.

3. Heterotrophs

This group of organisms are largely responsible for oxidizing SO in aerobic soils. It includes a wide range of bacteria, fungi and some actinobacteria.

Bacteria: *Arthrobacter*, *Bacillus*, *Micrococcus*, *Mycobacterium*, *Pseudomonas*

Fungi:*Alternaria*, *Fusarium*, *Trichoderma,Penicillium*

Sulphuric acid produced during oxidation of sulphur and H: S ratio is of great significance in reducing the pH of alkaline soils and in controlling potato scab and rot diseases caused by *Streptomyces*. The formation of sulphate / sulphuric acid is beneficial in agriculture in different ways : i) as it is the anion of strong mineral acid ($H_2 SO_4$) can render alkali soils fit for cultivation by correcting soil pH. ii) solubilize inorganic salts containing plant nutrients and thereby increase the level of soluble phosphate, potassium, calcium, magnesium etc. for plant nutrition.

4. Reduction of sulphate

Sulphate in the soil is assimilated by plants and microorganisms and incorporated into proteins. This is known as "assimilatory sulphate reduction". Sulphate can be reduced to hydrogen sulphide (H_2S) by sulphate reducing bacteria (eg. *Desulfovibrio* and *Desulfatomaculum)* and may diminish the availability of sulphur for plant nutrition. This is "dissimilatory sulphate reduction" which is not at all desirable from soil fertility and agricultural productivity view point.

Dissimilatory sulphate-reduction is favored by the alkaline and anaerobic conditions of soil and sulphates are reduced to hydrogen sulphide. For example, calcium sulphate is attacked under anaerobic condition by the members of the genus *Desulfovibrio* and *Desulfatomaculum* to release $H_2 S$.

$$CaSO_4 + 4H_2 \longrightarrow Ca(OH)_2 + H_2S + H_2O.$$

Hydrogen sulphide produced by the reduction of sulphate and sulphur containing amino acids decomposition is further oxidized by some species of green and purple phototrophic bacteria *(eg. Chlorobium, Chromatium)* to release elemental sulphur.

$$CO_2 + 2H_2 + H_2S \xrightarrow[Light]{} (CH_2O) + H_2O + 2S$$

The predominant sulphate-reducing bacterial genera in soil are *Desulfovibrio, Desulfatomaculum* and *Desulfomonas* (All are obligate anaerobes). Amongst these species *Desulfovibrio desulfuricans* are most ubiquitous, non-spore forming, obligate anaerobes that reduce sulphates at rapid rate in waterlogged / flooded soils. While species of *Desulfatomaculum* are spore forming, thermophilic obligate anaerobes that reduce sulphates in dry land soils. All sulphate-reducing bacteria excrete an enzyme called "desulfurases" or "bisulphate reductase". Rate of sulphate reduction in nature is enhanced by increasing water levels (flooding), high organic matter content and increased temperature.

14

Potassium Transformations in Soil

Potassium (K) is among the most important essential macronutrients for plant growth. About 98% of the potassium in the earth's crust exists in insoluble forms as rocks and silicate minerals, resulting in very low concentrations of soluble potassium in the soil for plant growth and development. Potassium is never incorporated in any organic compound and even in the plant system it remains as K^+ ion. It does not form any gas that could be lost to the atmosphere. It does not cause eutrophication. When left in the ecosystem as such, it is not at all toxic.Potassium activates enzymes, maintains cell turgor, enhances photosynthesis, reduces respiration, helps in transport of sugars and starches, and is essential for protein synthesis. In addition to plant metabolism, potassium improves crop quality because it helps in grain filling and kernel weight, strengthens straw, increases disease resistance and helps the plant better to withstand stress.

Potassium in Soils

Soils usually contain potassium in larger amounts than any other nutrients. Total K content in soil is 30,000- 50,000 kg/ha in top layer of the soil. Nearly 98 % of K bound in the mineral form and 2 % is in soil solution. Generally, potassium in soils exist in four forms (often called fractions), their availability to crop roots is considerably different.

1. Mineral K
2. Non exchangeable K
3. Exchangeable K
4. Soil solution K

1. Mineral K

It is an unavailable form. Also, this is the major natural source of potassium in soil. It is found on crystal structures of primary minerals like

Micas: Micas are 2:1 layer silicates where K^+ resides mainly between the layer silicates.

Muscovite: $KAl_2(AlSi_3O_{10})(FOH)_2$

Biotite: $K(MgFe)_3AlSi_3O_{10}(FOH)_2$

Phlogopite $KMg_3AlSi_3O_{10}(FOH)_2$

Feldspars: Feldspars are 3-dimensional crystal structure with K located throughout the mineral lattice. They are the largest natural potassium reserve.

Microcline : $KAlSi_3O_8$

The potassium from these sources can be released only through weathering. Thus, they supply very small quantities of potassium during a given growing season. Only their cumulative release over a period of years is of some importance which is enhanced by the solvent action of carbonic acid, stronger organic acid, inorganic acids, acidic clays and humus. Order of primary minerals in the ease of weathering is as follows. Biotite > Muscovite > K-feldspars > Phlogopite. The release of K from these minerals is also geared by the dissolution of the mineral from the roots of certain plants like elephant grass.

2. Non exchangeable K

As these K^+ ions cannot be replaced by ordinary exchange processes, it is called as non exchangeable. It is also known as slowly available form. It is the fixed K where K^+ ions are held by electrostatic forces as the integral part of secondary clay minerals like vermiculite, smectite, illite, chlorite,etc.These are not always immediately available form but contribute significantly to the maintenance of labile pool of K. It's release is too slow to meet crop demand during the growing season but is an extremely important reservoir of slowly available potassium.The release of this form of K occurs in response to depletion of readily available form either by plant uptake or leaching through the mechanism of cation exchange and diffusion.

Readily available potassium exists as $K+$ in the soil and is the form that constitutes the labile pool of this nutrient element for the plant. It has two forms*viz.*, Exchangeable K and Soil solution K

3. Exchangeable K

The K adsorbed on soil clay complex and replaceable with neutral salts in relatively short time and they are available to plant roots.

4. Soil solution K

It is the K+ ion present freely in the soil water. Higher values are found in arid or saline soils but much lower in soils of humid regions. Soil solution potassium is the most readily available form, preferably absorbed by the higher plants.

K cycle in soil

The potassium cycle is almost entirely inorganic. The major role of K in living organisms is the osmotic control. It is taken up, retained and excreted in ionic form (K^+). The amount of K in soil solution is relatively small but is in near equilibrium with the much larger amount of exchangeable K from which it is replenished. Soils also contain K in more slowly exchangeable forms (fixed K) which act as sources for crops. K present in clay minerals becomes available as these minerals weather. The immediate source of K for plants is the small amount which is in the soil solution. The concentration of potassium in the soil solution is of the range from 1 to 2%. When K in soil solution and exchangeable form are removed by plants uptake, the equilibrium is disturbed, K in the non-exchangeable and soil mineral fraction will be drawn upon.

Postassium cycle

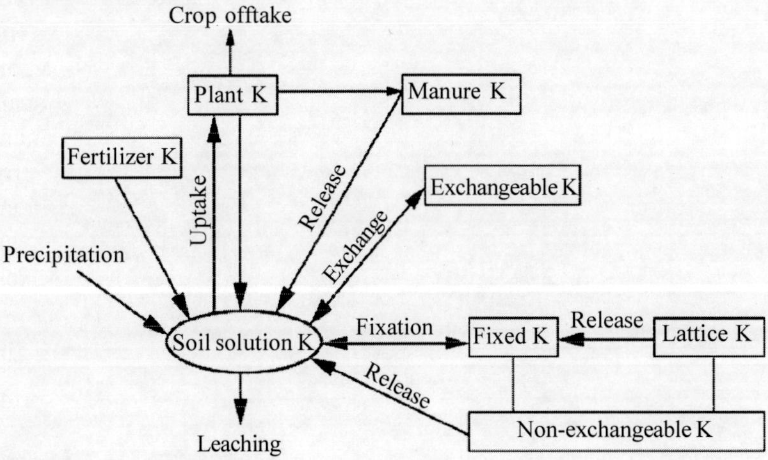

Mechanism of K release

Potassium is solubilized by production of inorganic and organic acids (citric, oxalic, alpha keto glutaric acids). production of mucilaginous capsules containing of exopolysaccharides by *Bacillus*, *Clostridium* and *Thiobacillus*. These organic acids produced enhance the dissolution of potassium compounds by supplying protons and by complexing Ca^{2+} ions. Microbially produced organic ligands include metabolic by products, extracellular enzymes, chelates and both simple and complex organic acids enhance the dissolution.

Microbes in K Transformation

Microbes like *Bacillus extroquens* and *Clostridium pasteurianum, Bacillus mucilogenoses, Frateuria aurentia* were found to dissolve potassium, silicates and aluminium from insoluble minerals. Fungal species like *Aspergillus niger, Pencillium frequentans* and *Cladosporium* were found to grow on muscovite, biotite and mica *in vitro* and also release K.

15

Zinc and Silica Transformations

Zinc is found in earth's crust to the tune of 0.008%. However about 50% of Indian soils exhibit deficiency of zinc with a content much below the critical level of 1.5 ppm. Soil contains both available and unavailable form of zinc. Water soluble, exchangeable, organically complexed and armorphous iron oxide bound forms are considered available zinc and crystalline iron oxide bound and residual zinc are considered as unavailable. The available fraction constitutes around 8% of the total zinc which is too low for utilization by crops in zinc deficient soils.

Zinc is required by crops in microquantities for its growth. The use of high analysis fertilizer due to cultivation of high yielding varieties, enhanced cropping intensity and the decline in the use of organic manures result in the deficiency of trace elements particularly zinc. With the liberal application of nitrogen the demand for zinc increases producing deficiency symptoms. The critical limit of zinc in dry matter for different crops and levels less than 20 ppm suggests the probability of zinc deficiency and less than 15 ppm indicate sure deficiency.

Zinc is a limiting factor in crop production and application of $ZnSO_4$ is preferred because of its water soluble nature. The zinc sulphate, when applied soil, gets transformed into different forms like Zn (OH), zinc carbonate in calcium rich alkali soils ; zinc phosphate in near neutral to alkali soils of high phosphorus application; and zinc sulphide under reduced conditions.

Solubilization of Zinc

Bacteria also have a zinc requirement and to absorb this, they act on soil zinc rendering it to a soluble from which is also absorbed by the plants.Zn solubilizing microorganisms are capable of solubilizing insoluble zinc compounds through the production of gluconic acid. Inoculation of zinc solubilizing bacteria (ZSB) either alone in soils inherently rich in native zinc or along with cheaper insoluble zinc compounds, like ZnO, ZnS and $ZnCO_3$ will be useful for enhancing Zn nutrition of crops. Inoculation of these organisms in zinc deficient soils will be helpful for the alleviation of zinc deficiency. Some of the predominant zinc solubilizers are *Bacillus cereus, Bacillus sphaericus, Bacillus subtilis, Gluconacetobacter diazotrophicus* and *Pseudomonas* sp.

Importance of Zinc solubilizing microorganisms

- They alter the Zn availability in soil upon application.
- They help in adsorption and accumulation of Zn in soil.
- They enhance the Zn concentration in the rhizosphere through organic acid production.
- They solubilize the insoluble compounds like, ZnO, Zn CO_3, ZnS and Spharelite ore.

Transformation of Silica

Silica is an important nutrient utilized by plants, trees, and grasses in the terrestrial biosphere. Silicate is transported by rivers and can be deposited in soils in the form of various siliceous polymorphs. Silica is generally found in soils as silicic acid or its ionic form at concentrations ranging from 0.1 to 0.6 mM. About 30 - 35 number of silicate minerals are found in soils (Mostly of K containing minerals- biotite and muscovite). Plants can readily uptake silicate in the form of silicic acid (H_4SiO_4) for the formation of phytoliths. Phytoliths are tiny rigid structures found within plant cells that aid in the structural integrity of the plant.

In soil, silica occurs in an unavailable polymerized state and unless it is deploymerized, is not available for plant absorption. This is made available by the biological action of roots and microbes and also by the physico chemical action like weathering. Among the plants the silica (SiO_2) concentrations are found to be higher in monocotyledons than in dicotyledons and its level increased from legumes < fruit crops < vegetables < grasses < grain crops. Grasses accumulates 4-20% silica in straw and almost every part of rice contains silica the requirement of which increases with the addition of more of nitrogen. Plants that have a Si concentration over 1% and a Si/Ca ratio >1 are called as 'Accumulators' and Excluders' have a Si concentration below 0.5% and a Si/ Ca ratio <0.5. Plants that do not meet these criteria are called 'intermediates'. Nitrogen addition makes the plant tender, and succulent, rendering a high demand for silica which reverses the nitrogen effect by making the plants sturdy and erect.Silica does not form a constituent of cellular component but primarily deposited on the walls of epidermis and vascular tissues conferring strength, rigidity and resistance to pests and diseases. Silica tends to maintain erectness of rice leaves and culms thereby increasing the photosynthesis.

Silica cycle

Silicon (Si), the second most abundant element (27%), usually occurs in the unavailable forms of silicates (SiO_3) of aluminum, magnesium, calcium, sodium, potassium, or iron. It's accessibility to the plant roots is generally governed by chemical and biological reactions of soil. Plants assimilate Si as soluble monosilicic acid resulting in strengthening of cell wall through various mechanisms. Strengthened cell wall due to higher accumulation of silicon is known to improve plant resistance to diseases, insect attack, and adverse climatic conditions in various plant species like rice, oat, barley, wheat, cucumber and sugarcane.

Plants absorb silica as monosilicic acid and this could arise in soil solution only when deploymerization of soil silicates occur. Organic weathering residues arising out of decomposition of residues, root exudates and / or metabolites from rhizosphere and non-rhizosphere microorganisms might result in the breakdown of silicate minerals releasing the ionic component. In soils, silicate minerals undergo chemical and physical weathering, resulting in the release of Si in solution, which is either combined with other elements to form clay minerals, or released towards the streams and the oceans or absorbed by the vegetation. During litter decomposition, plant Si is re-injected into the soil and can constitute an important pool of Si in soil. Crops can take up Si at a much faster rate, 300 kg ha"^{1} year"^{1} by sugar cane and 500 kg ha"^{1} year"^{1} by rice. Non-reincorporation of the straw to the field may lead to a depletion of plant-available Si in soils with a decline of cereal yields.

Solubilization of silica

Bacteria solubilizing silicates occur in the soil environment. The natural occurrence of silicate solubilizing bacteria (SSB) in rice ecosystem, in agro inputs and in materials of anthropogenic activities was observed. Several bacteria can solubilize the silica from minerals as well as from plant residues. *Bacillus* sp. isolated from granite crusher yard was found to solubilize various types of silicates minerals more effectively under *in vitro* conditions and also in soils of different types. Production of organic acids, exopolysaccharides, H_2S production and CO_2 production enable the bio dissolution of silicon. Organic acids play major role in weathering of Alumino silicate minerals. These bacteria can colonize mineral surfaces, organic acids can supply H^+ ions and promote hydrolysis on one hand and make complexes with cations on other hand. Citric, oxalic, keto, hydroxyl and carboxylic acids can from complexes with cations and promote their removal. Biodissolution of Si also releases K from minerals simultaneously. Some of the predominant silicate solubilizers are *Paenibacillus muciloginosus, Achromobacter xylosoxidans, Bacillus altitutinis* and *Pseudomonas gessardii.*

16

Bioconversion of Agricultural Wastes

Bioconversion refers to the recycling and reprocessing of organic waste materials using the system of biological decomposition or biodegradation into value added products. Composting is one way of conversion of organic waste into manures. It is defined as a method of solid waste management whereby the organic component of the solid waste stream is biologically decomposed under controlled conditions to a state in which it can be handled, stored or applied to the land without adversely affecting the environment.

Composting provides a partial solution to an issue of great concern in many communities. All around the world, landfills are filling up, garbage incineration is becoming increasingly unpopular, and other waste disposal options are becoming ever harder to find. Composting provides a way not only of reducing the amount of waste that needs to be disposed of, but also of converting it into a product that is useful for agriculture. Composting reduces waste, makes us less dependent on landfills, and decreases greenhouse gas emissions. Composting recycles essential nutrients back into the soil and promotes a prolific soil ecosystem.

Composting

Composting is the decomposition of organic matter by microorganisms under controlled conditions. It is a process of value addition and an eco friendly way of solid and liquid waste disposal / utilization. In other words, biological conversion of organic substrates under controlled conditions, into a hygienic, humus rich product, that conditions soil and nourish plants. Raw organic materials such as crop residues, animal wastes, food garbage, some municipal wastes and suitable industrial wastes, enhance their suitability for application to the soil as a fertilizing resource, after having undergone composting.

Substrates for composting

Most organic wastes can be successfully composted. Municipal solid wastes, biosolids, animal manures, agricultural, crop and food processing wastes are the excellent candidates for composting. Because they contain carbohydrates, proteins and lipids that the micro organisms can easily degrade.

Agricultural wastes

Crop residues like rice straw, wheat straw, sorghum straw, maize stover, cob, sunflower stalk, sugarcane trash, leaf litter etc., are available in large quantities in the agricultural sector. The agro industry based by products like pressmud, bagasse, coir pith, cotton waste, tea waste, mushroom spent waste can be processed to make them as manure. According to the type of wastes, the composting process can be adjusted and nutrient rich compost can be prepared.

Principles of Composting

Aerobic

- Sugars $(CH_2O)+O_2 \rightarrow CO_2+H_2O+$Energy
- Protein (Organic N) $\rightarrow NH_3 \rightarrow NO_2^- \rightarrow NO_3^-$
- Organic Sulphur + O $\rightarrow SO_4^-$
- Organic Phosphate $\rightarrow H_3PO_4 \rightarrow (HPO_4)_2$

Anaerobic composting

- CH_2O (Carbohydrates) $\rightarrow CH_3COOH$ (Acetic acid)
- $CH_3COOH \rightarrow CH_4 + CO_2$ (Methane)
- Organic Nitrogen $\rightarrow NH_3$ (Ammonia)
- Organic P $\rightarrow$ Reduced P

Properties of compostable wastes

Certain physical and chemical properties of the organic waste affect the composting process as well as characteristics of the finished product.

Chemical properties

Chemical characteristics may be considered in terms of nutrient quality and quantity. The relative quantity of C, N, P, S and other nutrients is important, but the quality of the substance that determines rate of decomposition. Lignin undergoes decomposition much lower than that of cellulose, the quality of lignin as a C source is much lower than that of cellulose. Maximum rates of decomposition are observed when substrates contain optimum available C, N, P and S ratios. Narrow C/N ratio can lead to loss of N from compost (through ammonia volatilization). Wider C/N ratio (> 40) in substrates such as oat straw promote immobilization of available N in the compost, resulting in a slowing of decomposition processes due to limited N availability. Addition of exogenous sources of N can overcome such deficiencies and prevent slowing of the composting process.

Physical properties

Most important physical properties of the substrate in relation to composting are particle size and moisture content.Particle size affects the movement of oxygen into the compost pile and microbial access to substrates. Large particle size promotes oxygen diffusion, but also minimizes the specific surface area of the substrate. This means that most of the substrates are not immediately accessible to microbes. Since efficient composting requires access to oxygen and to nutrients in the particles, it is necessary to balance these two requirements.Optimum moisture content for composting is generally between 40 and 60%. Too much moisture interferes with oxygen accessibility and slows composting. Too little moisture hinders diffusion of the soluble molecules and slows down the composting.

Microbial inoculants for composting

Naturally occurring organic materials are abundantly colonized with microorganisms to bring about decomposition. But the rate of decomposition will be slower. To hasten the decomposition as well as to increase the rate of decomposition, inoculation is considered essential.The inocula of selected species of fungi such as *Aspergillus, Penicillium, Trichoderma, Paecilomyces, Pleurotus* are found to play efficient role in promoting composting process and reduce the composting period by about four weeks.

Commonly used organisms for decomposing various substrates

Lignin rich waste	-	*Phanerochaete chrysosporium* *Tramates versicolor*
Lignocellulosic waste	-	*Pleurotus sajorcaju, Streptomyces*
Cellulosic waste	-	*Cellulomonas, Trichoderma*

The Science of composting

During the composting the microorganisms break down organic matter and produce carbon dioxide, water, heat, and humus, the relatively stable organic end product. Under optimal conditions the composting temperature changes during the process and three phases can be distinguished. 1) the mesophilic phase (moderate temperature phase), with a duration of a couple of days; 2) the thermophilic phase (high temperature phase), with duration from some days until several months; and finally3) a several-months cooling and maturation phase

1. Mesophilic Phase - I (10-40°C)

The first phase of the composting process starts with the degradation of the most easily degrading organics (sugars, organic acids, amino acids). It is mediated by aerobic microorganisms with consumption of oxygen and release of carbon dioxide and energy. It lasts only few days. An explosive growth of mesophilic bacteria and fungi is observed.

2. Thermophilic Phase (>40°C)

When the easily degradable compounds that are metabolized during the first stage are exhausted, the composting process continues with the more complex organic molecules; that is why the time for degradation is longer. During the thermophilic phase, high temperatures accelerate the breakdown of proteins, fats, and complex carbohydrates like cellulose and hemicellulose, the major structural molecules in plants. When these compounds are exhausted, the compost temperature gradually decreases and mesophilic microorganisms once again take part for the final phase of "curing" or maturation of the remaining organic matter. This phase can last from few days to several months depend on size of system.

3. Mesophilic Phase -II (10-40°C) - Maturation and Curing Phase

During the mesophilic phase II, actinomycetes appear. They play a basic role in the humus formation and produce aromatic compounds, which give to the end product a specific smell of wood soil.The maturation stage is characterized also with disintegration of the compost material: formation of small particles by a lot of invertebrates (earthworms, ticks, and centipede). The humus formation is strongly related to chemical and biological transformation of animal and vegetable wastes and with microbial synthesis. The humufication is attributed to oxidative polymerization of phenolics obtained by the catabolism of lignin, tannins, and polyphenols, or to new microbial biosynthesis. It can last several months.

Microbial characteristics of the composting process

Most of the wastes carry an indigenous population of microbes from the environment. Three major groups of soil microorganism are normally present when the composting process begins. Microbial population changes during the composting process progressing from mesophilic stage into thermophilic stage and then through a gradual cooling period, the final stabilization or curing stage.

Mesophilic microbiota initiates the decomposition of organic substrates and rapidly oxidizes the readily available substrates such as proteins, sugars and starch. Due to microbial metabolism, internal temperature may rise rapidly. The role of mesophile is a preparatory one. They bring the compost into a temperature range suitable for thermophiles by rapidly decomposing the readily available substrates. In addition to bacteria and fungi, soil animals such as earthworms and arthropods contribute to the decomposition process by reducing the size of particles greatly increasing the surface area to volume ratio.

Enzymatic activities during composting

Microbes in the compost pile cannot directly metabolize the insoluble particles of organic matter. Rather they produce hydrolytic extracellular enzymes to depolymerize the larger compounds to small fragments that are water soluble. Cellulase activity decreases the amount of cellulose present by about 25% in about 3 weeks. Lipases, proteases and amylase activities rise and fall during the successive stage of composting.

Composting Methods

- Coimbatore method
- Indore method
- Bangalore method

Coimbatore method

Composting is done in pits of different sizes depending on the waste material available. A layer of waste materials is first laid in the pit.It is moistened with a suspension of 5-10 kg cow dung in 2.5 to 5.0 lt. of water and 0.5 to 1.0 kg fine bone meal sprinkled over it uniformly. Similar layers are laid one over the other till the material rises 0.75 m above the ground level. It is finally plastered with wet mud and left undisturbed for 8 to 10 weeks. Plaster is then removed, material moistened with water, given a turning and made into a rectangular heap under a shade. It is left undisturbed till its use.

Indore Method

In the Indore method of composting, organic wastes are spread in the cattle shed to serve as bedding. Urine soaked material along with dung is removed every day and formed into a layer of about 15 cm thick at suitable sites. Urine soaked earth, scraped from cattle sheds is mixed with water and sprinkled over the layer of wastes twice or thrice a day. Layering process continued for about a fortnight. A thin layer of well decomposed compost is sprinkled over top and the heap given a turning and reformed. Old compost acts as inoculum for decomposing the material. The heap is left undisturbed for about a month. Then it is thoroughly moistened and given a turning. The compost is ready for application in another month.

Bangalore method

In the Bangalore method of composting, dry waste material of 25 cm thick is spread in a pit and a thick suspension of cow dung in water is sprinkled over for moistening. A thin layer of dry waste is laid over the moistened layer. The pit is filled alternately with dry layers of material and cow dung suspension till it rises 0.5 m above ground level. It is left exposed without covering for 15 days. It is given a turning, plastered with wet mud and left undisturbed for about 5 months or till required.

Agricultural Wastes Composting

Crop residues are the non-economic plant parts that are left in the field after harvest. The harvest refuses include straws, stubble, stover and haulms of different crops. Crop remains are also from thrashing sheds or that are discarded during crop processing. This includes process wastes like groundnut shell, oil cakes, rice husks and cobs of maize, sorghum and cumbu. The greatest potential as a biomass resource appears to be from the field residues of sorghum, maize, soybean, cotton, sugarcane etc. The following steps are involved in compost making.

i. Waste collection

Crop residues accumulated in different locations are to be brought to compost yard. The compost yard is located in any one corner of the farm with accessibility via good road. Water resource should also be available in sufficient quantity. The crop residues that are brought to compost yard should be heaped in one corner for further processing.

ii. Shredding of waste materials

Particle size is one of the factors that influence the composting. It is advisable to shred all the crop residues that are used for composting. Shredding the waste manually is labour intensive. Shredder machine can be employed to shred all the crop residue biomass. Particle size of 2 to 2.5 cm is recommended for quick composting.

iii. Mixing of green waste and brown waste

Carbon and nitrogen ratio decides the initiation of composting process. If C:N ratio is wide (100:1) composting will not take place. Narrow C:N ratio of 30:1 is ideal for composting. To get a narrow C:N ratio, carbon and nitrogen rich material should be mixed together. Green coloured waste materials like glyricidia leaves, parthenium, freshly harvested weeds and sesbania leaves are rich in nitrogen, whereas brown coloured waste material like straw, coir dust, dried leaves and dried grasses are rich in carbon. In any composting process these carbon and nitrogen rich material is to be mixed together to make the composting quicker rather than putting green waste alone or brown color waste alone for composting. Animal dung is also a good source of nitrogen. While making heap formation, alternative layers of carbon rich material, animal dung and nitrogen rich material are to be heaped to get a quicker result in composting.

iv. Compost heap formation

Minimum 4 feet height should be maintained for composting. The composting area should be elevated one and have sufficient shade. While heap formation, all the crop residues should be mixed together to form a heterogeneous material rather than a single homogenous material. Alternate layers of carbon and nitrogen rich material with intermittent layers of animal dung are essential. After heap formation the material should be thoroughly moistened.

v. Aerating the compost material

Sufficient quantity of oxygen should be available inside the compost heap. For this external air should be freely get in and comes out of the material. Normally to allow the fresh air to get inside, the compost heap should be turned upside down, once in fifteen days. In this process top layer comes to bottom and bottom layer goes to top. This process also activates the microbial process and compost process is hastened. In some cases air ventilating pipe maybe inserted vertically and horizontally, to allow the air to pass through. The wood chip that is available as waste in wood processing industry may also be used as bulking agent in the composting process. This bulking agent gives more air space to the compost material.

vi. Moisture maintenance

Throughout the composting period 60% moisture should be maintained. On any situation, compost material should not be allowed to dry. If the material becomes dry, all the microorganisms present in the crop residues will die and the compost process gets affected.

vii. Maturity

Volume reduction, black colour, earthy odor, reduction in particle size are all the physical factors to be observed for compost maturity. After satisfying with the compost maturity index, the compost heap can be disturbed and spread on the floor for curing.

viii. Compost application

Organic manures are highly regarded as good source of material to maintain soil health and increasing soil organic carbon content. Organic manures cannot be equated with inorganic fertilizers. But organic manures deliver all the nutrients to the soil but with little quantity. For one hectare of land 5 tons of enriched biocompost is recommended. It can be used as basal application in the field before taking up planting work.

Availability of Microorganisms

Bacteria

Bacillus brewis, B. circulans, B. coagulans, B. subtilis

Fungi

Mucor, Chaetomium, Thermophilum, Penicillium, Aspergillus, Humicola

Actinomycetes

Nocardia, Streptomyces, Thermoactinomycetes vulgaris, Thermomonospora spp.

Characteristics of good compost

- Directly increase the productivity of the soil.
- Maintain the soil in good condition.
- Increase the fertility of the soil over a long period of time.

- Possess a high organic and nutrient content.
- Be free of impurities and contaminants.

Benefits of Compost

- Volume reduction of waste.
- Composting temperature kill pathogen, weed seeds and seeds.
- During composting number of wastes from several sources are blended together.
- Excellent soil conditioner
- Saleable product
- Improves manure handling
- Reduces the risk of pollution
- Pathogen reduction
- Additional revenue.
- Suppress plant diseases and pests.
- Reduce or eliminate the need for chemical fertilizers.
- Promote higher yields of agricultural crops.
- Facilitate reforestation, wetlands restoration, and habitat revitalization efforts by amending contaminated, compacted, and marginal soils.
- Cost-effectively remediate soils contaminated by hazardous waste.

Average nutrient content (%) of some common compost

Manure/compost	N	P_2O_5	K_2O
Farmyard manure	0.80	0.41	0.74
Mixed farm wastes	0.87	0.59	2.22
Mixed dry residues	0.90	0.45	1.95
Cotton stalks	1.61	0.48	3.38
Water hyacinth compost	2.00	1.00	2.23
Dry grass/weeds	1.90	0.55	1.09
Vermicompost	1.60	2.20	0.67

Compost enrichment

Farm compost is poor in P content (0.4-0.8 percent). Addition of P makes the compost more balanced, and supplies nutrient to microorganisms for their multiplication and faster decomposition. The addition of P also reduces N losses.

Application of superphosphate, bonemeal or phosphate rock

1 kg of superphosphate or bonemeal is applied over each layer of animal dung. Low-grade phosphate rock can also be used for this purpose.

Use of animal bones: These can be broken into small pieces, boiled with lime water and drained, and the residue applied to the pits. Even the addition of raw bones, broken into small pieces and added to the pit, improves the nutrient value of compost significantly.

Wood ash waste can also be added to increase the K content of compost.

Addition of N-fixing and P-solubilizing cultures: The quality of compost can be further improved by the secondary inoculation of *Azotobacter or Azospirillum* (N_2-fixers); and *Bacillus* or *Pseudomonas* sp. (P solubilizers). These organisms, in the form of culture broth or water suspension can be sprinkled when the decomposing material is turned after one month. By this time, the temperature of the compost has also stabilized at about 35 °C.

17

Biodegradation of Agricultural Residues and Chemicals

Pesticides are the chemical substances that kill pests and herbicides are the chemicals that kill weeds. In the context of soil, pests are fungi, bacteria, insects, worms, and nematodes etc. that cause damage to field crops. Thus, in broad sense pesticides are insecticides, fungicides, bactericides, herbicides and nematicides that are used to control or inhibit plant diseases and insect pests. Although wide-scale application of pesticides and herbicides is an essential part of augmenting crop yields; excessive use of these chemicals leads to the microbial imbalance, environmental pollution and health hazards. An ideal pesticide should have the ability to destroy target pest quickly and should be able to degrade non-toxic substances as quickly as possible.

The ultimate "sink" of the pesticides applied in agriculture is soil. Soil being the storehouse of multitudes of microbes, in quantity and quality, receives the chemicals in various forms and acts as a scavenger of harmful substances. The efficiency and the competence to handle the chemicals vary with the soil and its physical, chemical and biological characteristics.

Effects of pesticides

Pesticides reaching the soil in significant quantities have direct effect on soil microbiological aspects, which in turn influence plant growth. Some of the most important effects caused by the pesticides are:

1. Alterations in ecological balance of the soil microflora.
2. Continued application of large quantities of pesticides may cause everlasting changes in the soil microflora.
3. Adverse effect on soil fertility and crop productivity.
4. Inhibition of N_2 fixing soil microorganisms such as *Rhizobium, Azotobacter, Azospirillum* etc. and cellulolytic and phosphate solubilizing microorganisms.

5. Suppression of nitrifying bacteria, *Nitrosomonas* and *Nitrobacter* by soil fumigants like ethylene and bromide.
6. Alterations in nitrogen balance of the soil.
7. Interference with ammonification in soil.
8. Adverse effect on mycorrhizal symbioses in plants and nodulation in legumes.
9. Alterations in the rhizosphere microflora, both quantitatively and qualitatively.

Persistence of pesticides in soil

How long an insecticide, fungicide, or herbicide persists in soil is of great importance in relation to pest management and environmental pollution. Persistence of pesticides in soil for longer period is undesirable because of the reasons: **a)** Accumulation of the chemicals in soil to highly toxic levels, **b)** May be assimilated by the plants and get accumulated in edible plant products, **c)** Accumulation in the edible portions of the root crops, **d)** To be get eroded with soil particles and may enter into the water streams, and finally leading to the soil, water and air pollutions. The effective persistence of pesticides in soil varies from a week to several years depending upon structure and properties of the constituents in the pesticide and availability of moisture in soil. For instance, the highly toxic phosphates do not persist for more than three months while chlorinated hydrocarbon insecticides (eg. DDT, aldrin, chlordane etc) are known to persist at least for 4-5 years and some times more than 15 years.

From the agricultural point of view, longer persistence of pesticides leading to accumulation of residues in soil may result into the increased absorption of such toxic chemicals by plants to the level at which the consumption of plant products may prove deleterious / hazardous to human beings as well as livestock's. There is a chronic problem of agricultural chemicals, having entered in food chain at highly inadmissible levels in several developing countries in the world. For example, intensive use of DDT to control insect pests and mercurial fungicides to control diseases in agriculture had been known to persist for longer period and thereby got accumulated in the food chain leading to food contamination and health hazards. Therefore, DDT and mercurial fungicides has been, banned to use in agriculture as well as in public health department.

1. Biodegradation of Pesticides in Soil

Pesticides reaching to the soil are acted upon by several physical, chemical, and biological forces. However, physical and chemical forces are acting upon/

degrading the pesticides to some extent, microorganism's plays major role in the degradation of pesticides. Many soil microorganisms have the ability to act upon pesticides and convert them into simpler non-toxic compounds. This process of degradation of pesticides and conversion into non-toxic compounds by microorganisms is known as "biodegradation". Not all pesticides reaching to the soil are biodegradable and such chemicals that show complete resistance to biodegradation are called "recalcitrant".

The chemical reactions leading to biodegradation of pesticides fall into several broad categories which are discussed in brief in the following paragraphs.

a) Detoxification

It is the conversion of the pesticide molecule to a non-toxic compound. Detoxification is not synonymous with degradation. Since a single change in the side chain of a complex molecule may turn into the chemical non-toxic.

b) Degradation

The breaking down / transformation of a complex substrate into simpler products leading finally to mineralization. Degradation is often considered to be synonymous with mineralization, e.g. Thiram (fungicide) is degraded by a strain of *Pseudomonas* and the degradation products are dimethylamine, proteins, sulpholipids, etc.

c) Conjugation (complex formation or addition reaction)

In which an organism makes the substrate more complex or combines the pesticide with cell metabolites. Conjugation or the formation of addition product is accomplished by those organisms catalyzing the reaction of addition of an amino acid, organic acid or methyl crown to the substrate, for e.g., in the microbial metabolism of sodium dimethyl dithiocarbamate, the organism combines the fungicide with an amino acid molecule normally present in the cell and thereby inactivate the pesticides/chemical.

d) Activation

It is the conversion of non-toxic substrate into a toxic molecule, for eg. Herbicide, 4-butyric acid (2, 4-D B) and the insecticide Phorate are transformed and activated microbiologically in soil to give metabolites that are toxic to weeds and insects.

e) Changing the spectrum of toxicity

Some fungicides/pesticides are designed to control one particular group of organisms / pests, but they are metabolized to yield products inhibitory to entirely dissimilar groups of organisms, for e.g. the fungicide,Pentachloronitrobenzene (PCNB) fungicide is converted in soil to chlorinated benzoic acids that kill plants.

Biodegradation of pesticides / herbicides is greatly influenced by the soil factors like moisture, temperature, pH and organic matter content, in addition to microbial population and pesticide solubility. Optimum temperature, moisture and organic matter in soil provide congenial environment for the break down or retention of any pesticide added in the soil. Most of the organic pesticides degrade within a short period (3-6 months) under tropical conditions. Metabolic activities of bacteria, fungi and actinomycetes have the significant role in the degradation of pesticides.

Criteria for Biodegradation

For successful biodegradation of pesticide in soil, following aspects must be taken into consideration. i) Organisms must have necessary catabolic activity required for degradation of contaminant at fast rate to bring down the concentration of contaminant, ii) the target contaminant must be bioavailability, iii) soil conditions must be congenial for microbial /plant growth and enzymatic activity and iv) cost of bioremediation must be less than other technologies of removal of contaminants.

Strategies for Bioremediation: For the successful biodegradation, following strategies are needed.

a) Passive/ intrinsic bioremediation: It is the natural bioremediation of contaminant by the indigenous microorganisms but the rate of degradation is very slow.

b) Biostimulation: Practice of addition of nitrogen and phosphorus to stimulate indigenous microorganisms in soil.

c) Bioventing: Process/way of biostimulation by which gases stimulants like oxygen and methane are added or forced into soil to stimulate microbial activity.

d) Bioaugmentation: It is the inoculation/introduction of microorganisms in the contaminated site/soil to facilitate biodegradation.

e) Composting: Piles of contaminated soils are constructed and treated with aerobic thermophilic microorganisms to degrade contaminants. Periodic physical mixing and moistening of piles are done to promote microbial activity.

f) Phytoremediation: Can be achieved directly by planting plants which hyperaccumulate heavy metals or indirectly stimulating microorganisms in the rhizosphere.

g) Bioremediation: Process of detoxification of toxic/unwanted chemicals / contaminants in the soil and other environment by using microorganisms.

h) Mineralization: Complete conversion of an organic contaminant to its inorganic constituent by a species or group of microorganisms.

Bioremediation

Introduction

In the twentieth century, the ever increase in the global human population and industrialization led to the exploitation of natural resources. The increased usage of heavy metals and its disposal in the ecosystems around the world and also in India is becoming an important ecological issue. This indiscriminate release of heavy metals into the ecosystem has already posing air, water and soil pollution causing various, deleterious and fatal effects on humans and the stability of the ecosystems. As a result of this, several living forms which are better suited to the polluted environment have out bursted causing ecological imbalance. Unlike soil, water body has its own ability to maintain its natural state through process called self purification. However when the discharge of pollutants is heavy, the process of self purification of water bodies is adversely affected and the water remains polluted. Unlike organic contaminants, heavy metals are not biologically degradable, and therefore can persist in the environment for a long duration. Some important heavy metals are Zinc (Zn), Chromium (Cr), Cadmium (Cd), Arsenic (Ar) etc.

Bioremediation can be defined as any process that uses microorganisms, green plant sor their enzymes to return the natural environment altered by contaminants to its original condition. Bioremediation may be employed to attack specific soil contaminants, such as degradation of chlorinated hydrocarbons by bacteria. An example of a more general approach is the cleanup of oil spills by the addition of nitrate and/or sulfate fertilizers to facilitate the decomposition of crude oil by indigenous or exogenous bacteria. Mycoremediation is a form of bioremediation that uses fungi to reduce the level of contamination in a given environment. Fungi can release specific enzymes and acids that break down the major components of plant fiber.

Bioremediation technologies can be generally classified as *in situ* or *ex situ*. In *situ* bioremediation involves treating the contaminated material at the site while *ex situ* involves the removal of the contaminated material to be treated elsewhere.

Some examples of bioremediation technologies are bioventing, landfarming, bioreactor, composting, bioaugumentation and biostimulation.

Bioremediation Strategies

The different techniques are employed depending on the degree of saturation and aeration of an area. *In situ* techniques are defined as those that are applied to soil and groundwater at the site with minimal disturbance. *Ex situ* techniques are those that are applied to soil and groundwater at the site which has been removed from the site via excavation (soil) or pumping (water). Bioaugmentation techniques involve the addition of microorganisms with the ability to degrade pollutants.

In situ bioremediation

These techniques are generally the most desirable options due to lower cost and less disturbance since they provide the treatment in place avoiding excavation and transport of contaminants. *In situ* treatment is limited by the depth of the soil that can be effectively treated. In many soils effective oxygen diffusion for desirable rates of bioremediation extend to a range of only a few centimeters to about 30 cm into the soil, although depths of 60 cm and greater have been effectively treated in some cases. The most important land treatments are incineration and excavation.

Bioventing

Bioventing is the most common *in situ* treatment and involves supplying air and nutrients through wells to contaminated soil to stimulate the indigenous bacteria. Bioventing employs low airflow rates and provides only the amount of oxygen necessary for the biodegradation while minimizing volatilization and release of contaminants to the atmosphere. It works for simple hydrocarbons and can be used where the contamination is deep under the surface.

In situ biodegradation involves supplying oxygen and nutrients by circulating aqueous solutions through contaminated soils to stimulate naturally occurring bacteria to degrade organic contaminants. It can be used for soil and groundwater. Generally, this technique includes conditions such as the infiltration of water-containing nutrients and oxygen or other electron acceptors for groundwater treatment.

Biosparging

Biosparging involves the injection of air under pressure below the water table to increase groundwater oxygen concentrations and enhance the rate of biological degradation of contaminants by naturally occurring bacteria. Biosparging increases the mixing in the saturated zone and thereby increases the contact between soil and groundwater. The ease and low cost of installing small-diameter air injection points allows considerable flexibility in the design and construction of the system.

Bioaugmentation

Bioremediation frequently involves the addition of microorganisms indigenous or exogenous to the contaminated sites. Two factors limit the use of added microbial cultures in a land treatment unit: 1) nonindigenous cultures rarely compete well enough with an indigenous population to develop and sustain useful population levels and 2) most soils with long-term exposure to biodegradable waste have indigenous microorganisms that are effective degrades if the land treatment unit is well managed.

Ex situ bioremediation

Ex Situ Bioremediation is the conversion of contaminants to non toxic end products. These techniques involve the excavation or removal of contaminated soil from ground.

Land farming

It is a simple technique in which contaminated soil is excavated and spread over a prepared bed and periodically tilled until pollutants are degraded. The goal is to stimulate indigenous biodegradative microorganisms and facilitate their aerobic degradation of contaminants. In general, the practice is limited to the treatment of superficial 10–35 cm of soil. Since landfarming has the potential to reduce monitoring and maintenance costs, as well as clean-up liabilities, it has received much attention as a disposal alternative.

Composting

It is a technique that involves combining contaminated soil with nonhazardous organic amendments such as manure or agricultural wastes. The presence of these organic materials supports the development of a rich microbial population and elevated temperature characteristic of composting.

Biopile

They are hybrid of landfarming and composting. Essentially, engineered cells are constructed as aerated composted piles. Typically used for treatment of surface contamination with petroleum hydrocarbons. They are a refined version of landfarming that tend to control physical losses of the contaminants by leaching and volatilization. Biopiles provide a favorable environment for indigenous aerobic and anaerobic microorganisms.

Bioreactors

Slurry reactors or aqueous reactors are used for *ex situ* treatment of contaminated soil and water pumped up from a contaminated plume. Bioremediation in reactors involves the processing of contaminated solid material (soil, sediment, sludge) or water through an engineered containment system. A slurry bioreactor may be defined as a containment vessel and apparatus used to create a three-phase (solid, liquid, and gas) mixing condition to increase the bioremediation rate of soil- bound and water-soluble pollutants as a water slurry of the contaminated soil and biomass (usually indigenous microorganisms) capable of degrading target contaminants. In general, the rate and extent of biodegradation are greater in a bioreactor system than *in situ* or in solid-phase systems because the contained environment is more manageable and hence more controllable and predictable. Despite the advantages of reactor systems, there are some disadvantages. The contaminated soil requires pre treatment (e.g., excavation) or alternatively the contaminant can be stripped from the soil via soil washing or physical extraction (e.g., vacuum extraction) before being placed in a bioreactor.

Microorganisms Involved in Bioremediation

Bioremediation involves various microorganisms that are able to degrade and reduce toxicity of environmental pollutants. Those microbes can be either naturally present in the site of bioremediation or isolated from other sites and inoculated artificially.

Bacteria

Bacteria are widely diverse organisms, and thus make excellent players in biodegradation and bioremediation.

i. *Pseudomonas putida*

Pseudomonas putida is a gram-negative soil bacterium that is involved in the bioremediation of toluene. It is also capable of degrading naphthalene, a product of petroleum refining, in contaminated soils.

ii. *Dechloromonas aromatica*

It is a rod-shaped bacterium which can oxidize aromatics including benzoate, chlorobenzoate, and toluene. It is the only organism able to oxidize benzene anaerobically. Due to the high propensity of benzene contamination, especially in ground and surface water, *D. aromatica* is especially useful for *in situ* bioremediation of this substance.

iii. *Deinococcus radiodurans*

It is a radiation-resistant extremophile bacterium that is genetically engineered for the bioremediation of solvents and heavy metals. It has been shown to degrade ionic mercury and toluene in radioactive mixed waste environments.

iv. *Methylibium petroleiphilum*

It is a bacterium capable of methyl tert-butyl ether (MTBE) bioremediation. It degrades MTBE by using the contaminant as the sole carbon and energy source.

Fungi

Fungi have significant roles in bioremediation because of their participation in the cycling of elements through decomposition and transformation of organic and inorganic materials. The most important feature of these fungi is their enzymatic functional ability to metabolize complex chemicals.

Phanerochaete chrysosporium

P.chrysosporium was the first fungi linked to degradation of organic pollutants. It has strong potential for bioremediation of pesticides, PAHs, dioxins, carbon tetrachloride, and many other pollutants.

Archaea

Few extreme halophilic strains of archaea belonging to genera *Halobacterium*, *Haloferax*, and *Halococcus* were found to biodegrade crude oil and hydrocarbons. They use various kinds of hydrocarbons *viz*., benzene, toluene, biphenyl, and naphthalene as their carbon or energy sources.

Advantages of bioremediation

- Bioremediation is a natural process and is therefore perceived by the public as an acceptable waste treatment process for contaminated soil. The residues of the treatment are usually harmless products that include carbon dioxide, water, and cell biomass.

- Bioremediation is useful for the complete destruction of a wide variety of contaminants. Many compounds that are legally considered to be hazardous can be transformed to harmless products.
- Instead of transferring contaminants from one environmental medium to another, for example, from land to water or air, the complete destruction of target pollutants is possible.
- Bioremediation can often be carried out on site, often without causing a major disruption of normal activities. This also eliminates the need to transport quantities of waste off site and the potential threats to human health and the environment that can arise during transportation.
- Bioremediation can prove less expensive than other technologies that are used for clean-up of hazardous waste.

Disadvantages of bioremediation

- Bioremediation is limited to those compounds that are biodegradable, susceptible to rapid and complete degradation.
- There are some concerns that the products of biodegradation may be more persistent or toxic than the parent compound.
- Biological processes are often highly specific. Important site factors required for success include the presence of metabolically capable microbial populations, suitable environmental growth conditions, and appropriate levels of nutrients and contaminants.
- It is difficult to extrapolate from bench and pilot-scale studies to full-scale field operations.
- Bioremediation often takes longer than other treatment options, such as excavation and removal of soil or incineration.

Evaluating performance of bioremediation is difficult, and there are no acceptable endpoints for bioremediation treatments.

18

Rhizosphere and Its Importance

Rhizosphere

The term Rhizosphere was introduced by the German scientist, Hiltner in 1904. It is region of the soil, which is subjected to the influence of plant roots. It is the unique environment under the influence of plant roots. The roots of plants are not only involved in the uptake of mineral nutrients and water for plant growth, but they also release a wide range of organic compounds in the surrounding soil.

Rhizosphere ecosystem

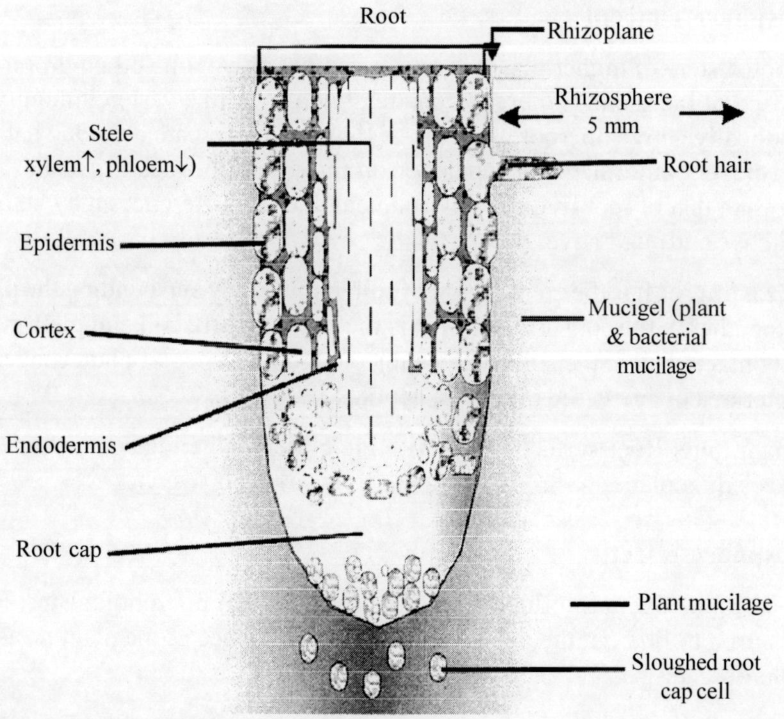

Consequently, the area of soil in contact with the plant root, termed the rhizosphere, is a site of intense microbial activity. Not surprisingly, many microorganisms are present at higher numbers on the surface of plant roots and in the rhizosphere than in soil not influenced by the presence of roots. Hence rhizosphere is characterized by greater microbiological activity than the soil away from plant roots.

The term rhizosphere is often divided into two general areas, the inner rhizosphere (very root surface) and the outer rhizosphere (immediate adjacent soil). The microbial numbers are larger in the inner zone, where the bio chemical interactions between microorganisms and roots are most pronounced.

The rhizosphere is a key soil habitat, where the numerous interactions taking place between plant root and soil microorganisms will determine growth conditions for both the plant and the microorganisms in the rhizosphere. Indeed, the rhizosphere microbiota exerts an important influence on roots and on growth of the plant. Consequently, the rhizosphere represents an important soil habitat where introduction of beneficial microorganisms (e.g. biofertilizers, phytostimulators, and biopesticides) as inoculants can result in significant improvements in crop yield and /or crop quality.

Rhizosphere concept

The root system of higher plants is associated not only with soil environment composed of inorganic and organic matter, but also with a vast community of metabolically active microorganisms. As living plants create a unique habitat around the roots, the microbial population on and around the roots is considerably higher than that of root free soil environment and the differences may be both quantitative and qualitative.

1. Rhizosphere: It is the zone/region of soil immediately surrounding the plant roots together with root surfaces, or it is the region where soil and plant roots make contact, or it is the soil region subjected to influence of plant roots and characterized by increased microbial actions.

2. Rhizoplane: Root surface along with the closely adhering soil particles is termed as rhizoplane.

Rhizosphere effect

It indicates the over all influence of plant roots on soil microorganisms. R: S ratio is used to find out the degree or extend of influence of the plant roots on soil microorganisms.

R:S ratio is defined as the ratio of microbial numbers per unit weight of rhizosphere (R), to the numbers in a unit weight of the adjacent non-rhizosphere soil (S). The rhizosphere effect is consistently greater for bacteria than for other microbes. The short, gram-negative rods, which almost invariably make up to a larger percent of the rhizosphere. On a generic basis, root effect appears to be more pronounced on *Pseudomonas*, *Flavobacterium*, *Alcaligenes, Arthrobacter*, *Brevibacterium, Cornebacterium, Micrococcus, Xanthomonas,* and *Bacillus.* Anaerobic bacteria are affected by the root; this may be attributed to the reduced O_2 tension resulting from root and microbial respiration. Bacterial density in the rhizosphere soil is 10^9 per gram and it may cover 4-10 % of the root area. Fast growers and metabolically active population are dominating in the rhizosphere region. This suggests that the rhizosphere flora has a greater ability to effect rapid biochemical changes than the organisms of fallow land. Selective influence of the root system is exhibited in few cases like preferential stimulation of G-ve, non spore forming short rods, vitamin requiring bacteria, abundance of N_2 fixing and P solubilizing organisms etc., The spectrum of genera varies with plant species, age and kind of the soil. Actinobacteria and algae are not significantly benefited by their proximity to roots.

Root Exudates

One of the most important factors responsible for rhizosphere effect is the great variety of organic substances available at the root region by the way of exudates from roots, which directly or indirectly influence the quality and quantity of the micro organisms in the root region. The substances exuded by plant roots include amino acids (asparagine,alanine, leucine, serine, glycine, lysin, prolein, trytophane), sugars (glucose, fructose, sucrose, xylose, arabinose, oligosaccharides), organic acids (tartaric, oxalic, malic, acetic, propionic, succinic, fumaric, butric), growth factors (biotin, thiamine, niacin, inositol, pyridoxin, p- amino benzoic acid), nucleotides, flavanones, enzymes and many other unidentified substances.

The nature and amount of substances exuded are dependent on the species of plant, age and environmental conditions under they grow. The products of photosynthesis are translocated to the root system in less than 12 hours and indicating the influence of metabolism of plants determining the rhizosphere effect. The root cap and ares of active growth are primary regions of root exudation.

Influence of the rhizosphere microflora on plant growth

The rhizosphere community has either a favourable or detrimental influence of plant development. Because the microflora is so intimately related with the root system, partially covering its surface, any beneficial or toxic substance produced can cause an immediate response. Vast microorganisms demand a variety of anions and cations for their own development and as a result immobilization take place. Aerobic organisms remove O_2 from the environment and add CO_2 and due to lowering of O_2 or increasing of CO_2 tension, it may reduce root elongation, development and reduce the water and nutrient uptake. It may however favour plant development by producing growth stimulating substances, forming of stable soil structure, releasing elements in organic forms through the mineralization of organic complexes, and by entering into symbiotic root associations.

Influence of the Plant on Microflora

The microflora is affected in many ways by the growing plant. The important contribution is the provision of excretion products and sloughed off tissues to serve as energy, carbon, nitrogen or growth factors. Due to the large quantities of CO_2 liberated by the root zone during respiration, the carbonic acid is formed. It solubilizes the unavailable elements to make available and that will be assimilated by the microflora.

Historical Background

The term "Rhizosphere" was introduced for the first time by the German scientist Hiltner (1904) to denote that region of soil which is subjected to the influence of plant roots. The concept of "Rhizosphere Phenomenon" which shows the mutual interaction of roots and microorganisms was came into existence with the work of Starkey et al (1929), Clark (1939) and Rauath and Katznelson (1957).

N. V. Krassinikov (1934) found that free living nitrogen fixing bacteria, *Azotobacter* were unable to grow in the wheat rhizosphere.

Starkey (1938) examined the rhizosphere region of some plant species and demonstrated the effect of root exudates on the predominance of bacterial population in particular and other soil microorganisms in general in the rhizosphere region. Thus, he put forth the concept of "Rhizosphere effect / phenomenon" for the first time.

F E Clark (1949) introduced / coined the term "Rhizoplane" to denote the root surface together with the closely adhering soil particles.

R.I. Perotti (1925) suggested the boundaries of the rhizosphere region and showed that it was bounded on one side by the general soil region (called as Edaphosphere) and on the other side by the root tissues (called Histosphere).

G. Graf and S. Poschenrieder (1930) divided the rhizosphere region into two general areas i.e. outer rhizosphere and inner rhizosphere for the purpose of describing the same site of microbial action.

H. Katznelson (1946) suggested the R:S ratio i.e. the ratio between the microbial population in the rhizosphere (R) and in the soil (S) to find out the degree or extent of plant roots effect on soil microorganisms. R: S ratio gives a good picture of the relative stimulation of the microorganisms in the rhizosphere of different plant species.

R: S **ratio** is defined as the ratio of microbial population per unit weight of rhizosphere soil (R), to the microbial population per unit weight of the adjacent non-rhizosphere soil (S).

A. G. Lochhead and H. Katznelson (1940) examined in detail the qualitative differences between the microflora of the rhizosphere and microflora of the non-rhizosphere region and reported that gram-negative, rod shaped and non-spore forming bacteria are abundant in the rhizosphere than in the non-rhizosphere soil.

C. Thom and H. Humfeld (1932) found that corn roots in acidic soils yielded predominantly *Trichoderma* while roots from alkaline soils mainly contained *Penicillium.*

M J. Timonin (1940) reported some differences in the fungal types and population in the rhizosphere of cereals and legumes. R: S ratio of fungal population was believed to be narrow in most of the plant species, usually not exceeding 10.

E. A. Peterson and others (1958) reported that the plant age and soil type influence the nature of fungal flora in the rhizosphere, and the number of fungal population gradually increases with the age of plant.

M. Adati (1932) studied many crops and found that though actinomycetes were relatively less stimulated than bacteria, but in some cases the R: S ratio of actinomycetes was as high as 62.

R. Venkatesan and G. Rangaswami (1965) studied the rhizosphere effect in rice plant on bacteria, actinomycetes and fungi and reported that (i) for actinornycetes R: S was more (ranging from 0 to 25) depending on the age of plant roots and the dominant genera reported were Nocardia, (ii) R:S ratio reduced with the depth of soil.

E. A. Gonsalves and V.S. Yalavigi (1960) reported the presence of greater number of algae in the rhizosphere.

J. W. Rouatt et al reported positive rhizosphere effect on protozoa, but a negative effect on algae in wheat plants.

Microorganisms in Rhizosphere and Rhizosphere Effect

The rhizosphere region is a highly favorable habitat for the proliferation, activity and metabolism of numerous microorganisms. The rhizosphere microflora can be enumerated intensively by microscopic, cultural and biochemical techniques. Microscopic techniques reveal the types of organisms present and their physical association with the outer root tissue surface / root hairs. The cultural technique most commonly followed is "serial dilution and plate count method" which reveal the quantitative and qualitative population of microflora. At the same time, a cultural method shows the selective enhancement of certain categories of bacteria. The biochemical techniques used are designed to measure a specific change brought about by the plant or by the microflora. The rhizosphere effect on most commonly found microorganisms *viz.* bacteria, actinobacteria, fungi, algae and protozoa is being discussed in the following paragraphs.

A. Bacteria

The greater rhizosphere effect is observed with bacteria (R: S values ranging from 10-20 or more) than with actinobacteria and fungi. Gram-negative, rod shaped, non-sporulating bacteria which respond to root exudates are predominant in the rhizosphere *(Pseudomonas, Agrobacterium).* While Gram-positive, rods, cocci and aerobic spore forming *(Bacillus, Clostridium)* are comparatively rare in the rhizosphere. The most common genera of bacteria are: *Pseudomonas, Arthrobacter, Agrobacterium, Alcaligenes, Azotobacter, Mycobacterium, Flavobacter, Cellulomonas, Micrococcus* and others have been reported to be either abundant or sparse in the rhizosphere. From the agronomic point of view, the abundance of nitrogen fixing and phosphate solubilizing bacteria in the rhizosphere assumes a great importance. The aerobic bacteria are relatively less in the rhizosphere because of the reduced oxygen levels due to root respiration. The bacterial population in the rhizosphere is enormous in the ranging form 108 to 109 per gram of rhizosphere soil. They cover about 4-10% of the total root area occurring profusely on the root hair region and rarely in the root tips. There is predominance of amino acids and growth factors required by bacteria, are readily provided by the root exudates in the region of rhizosphere.

B. Fungi

In contrast to their effects on bacteria, plant roots do not alter / enhance the total count of fungi in the rhizosphere. However, rhizosphere effect is selective and significant on specific fungal genera *(Fusarium, Verticillium, Aspergillus* and *Penicillium)* which are stimulated. The R:S ratio of fungal population is believed to be narrow in most of the plants, usually not exceeding to 10. The soil / serial dilution and plating technique used for the enumeration of rhizosphere fungi may often give erratic results as most of the spore formers produce abundant colonies in culture media giving a wrong picture / estimate *(eg Aspergilli* and *Penicillia).* In fact the mycelial forms are more dominant in the field. The zoospore / forming lower fungi such as *Phytophthora, Pythium, Aphanomyces* are strongly attracted to the roots in response to particular chemical compounds excreted by the roots and cause diseases under favorable conditions. Several fungi *eg. Gibberella fujikuroi* produces phytohormones and influence the plant growth.

C. Actinobacteria, Protozoa and Algae

Stimulation of actinobacteria in the rhizosphere has not been studied in much detail so far. It is generally understood that the actinomycetes are less stimulated in the rhizosphere than bacteria. However, when antagonistic actinobacteria increase in number they suppress bacteria. Actinobacteria may also increase in number when antibacterial agents are sprayed on the crop. Among the actinobacteria, the phosphate solublizers (eg. *Nocardia, Streptomyces)* have a dominant role to play.

As a rule actinobacteria, protozoa and algae are not significantly influenced by their proximity to the plant roots and their R: S ratios rarely exceed 2 to 3: 1 and around roots of plants, R: S ratio for these microorganisms may go to high. Because of large bacterial community, an increase in the number or activity of protozoa is expected in the rhizosphere. Flagellates and amoebae are dominant and ciliates are rare in the region.

Factors Affecting Microbial Flora of the Rhizosphere

The most important factors which affect / influence the microbial flora of the rhizosphere or rhizosphere effect are: soil type and its moisture, soil amendments, soil pH, proximity of root with soil, plant species, and age of plant and root exudates.

A. Soil type and its moisture: In general, microbial activity and population is high in the rhizosphere region of the plants grown in sandy soils and least in the high humus soils, and rhizosphere organisms are more when the soil moisture is low. Thus, the rhizosphere effect is more in the sandy soils with low moisture content.

B. Soil amendments and fertilizers: Crop residues, animal manure and chemical fertilizers applied to the soil cause no appreciable effect on the quantitative or qualitative differences in the microflora of rhizosphere. In general, the character of vegetation is more important than the fertility level of the soil.

C. Soil pH/ Rhizosphere pH: Respiration by the rhizosphere microflora may lead to the change in soil rhizosphere pH. If the activity and population of the rhizosphere microflora is more, then the pH of rhizosphere region is lower than that of surrounding soil or non-rhizosphere soil. Rhizosphere effect for bacteria and protozoa is more in slightly alkaline soil and for that of fungi is more in acidic soils.

D. Proximity of root with Soil: Soil samples taken progressively closer to the root system have increasingly greater population of bacteria, actinobacteria and decreases with the distance and depth from the root system. Rhizosphere effect decline sharply with increasing distance between plant root and soil.

E. Plant Species: Different plant species inhabit often somewhat variable microflora in the rhizosphere region. The qualitative and quantitative differences are attributed to variations in the rooting habits, tissue composition and excretion products. In general, legumes show / produce a more pronounced rhizosphere effect than grasses or cereals. Biennials, due to their long growth period exert more prolonged stimulation on rhizosphere effect than annuals.

F. Age of Plant: The age of plant also alter the rhizosphere microflora and the stage of plant maturity controls the magnitude of rhizosphere effect and degree of response to specific microorganisms. The rhizosphere microflora increases in number with the age of the plant and reaching at peak during flowering which is the most active period of plant growth and metabolism. Hence, the rhizosphere effect was found to be more at the time of flowering than in the seedling or full maturity stage of the plants. The fungal flora (especially, *Cellulolytic* and *Amylolytic)* of the rhizosphere usually increases even after fruiting and the onset of senescence due to accumulation of moribund tissue and sloughed off root parts / tissues: whereas, bacterial flora of the rhizosphere decreases after the flowering period and fruit setting.

G. Root / exudates /excretion: One of the most important factors responsible for rhizosphere effect is the availability of a great variety of organic substances at the root region by way of root exudates/excretions. The quantitative and qualitative differences in the microflora of the rhizosphere from that of general soil are mainly due to influences of root exudates. The spectrum of chemical composition root exudates varies widely, and hence their influence on the microflora also varies widely.

Root exudates are composed of the chemical substances like

S. No	Root exudates	Chemical Substances
1.	Amino Acids	All naturally occurring amino acids.
2.	Organic acids	Acetic, butyric, citric, fumaric, lactic, malic, propionic, succinic etc.
3.	Carbohydrates / sugars	Arabinose, fructose, galactose, glucose, maltose, mannose, oligosaccharides, raffinose, ribose, sucrose, xylose etc.
4.	Nucleic acid derivatives	Adenine, cystidine, guanine, uridine
5.	Growth factors (phytohormones)	Biotin, choline, inositol, pyridoxine etc
6.	Vitamins	Thiamine, nicotinic acid, biotin etc
7.	Enzymes	Amylase, invertase, protease, phosphatase etc.
8.	Other compounds	Auxins, glutamine, glycosides, hydrocyanic acid peptides, UV-absorbing compounds, nematode attracting factors, spore germination stimulators, spore inhibitors etc.

The nature and amount of chemical substances thus exuded are dependent on the species of plant, plant age, inorganic nutrients, temperature, light intercity, O_2 / CO_2 level, root injury etc. Another source of nutrients for the microorganisms in the rhizosphere region is the sloughed off root epidermis which exert selective stimulation effect on some specific groups of microorganisms. For instance, glucose and amino acids in the exudates readily attract gram-negative rods which predominantly colonize the roots. Sugars and amino acids in the root exudates stimulate the germination of chlamydospores and other resting spores of fungi; stimulation effect of root exudates on plant pathogenic fungi, nematodes is also well known.

Alterations in Rhizosphere Microflora

Foliar application of various chemicals leads to alterations in the rhizosphere microflora by changing the pattern of root exudates. The pattern of the rhizosphere microflora i.e. numbers and species composition can be changed / altered by various factors, such as: (i) Soil amendments, (ii) Foliar application of fertilizers / nutrients, fungicides, insecticides and hormones and (iii) Bacterization / microbial seed inoculants.

A. Soil amendments

Soil amendments with inorganic and organic fertilizers can alter the rhizosphere microflora and an understanding of the type of changes in the microflora can be useful in the indirect control of pathogens. Experimental results (1985) showed that amendment of soil with neem-cake can stimulate the activity of actinobacteria which results into the reduction of propagules of *Macrophomina phaseolina.* It is also known to control phytopathogenic nematodes in soil by stimulating nematode trapping fungi. Amendment of soil with castor and bean leaves stimulate the activity of *Trichoderma viride* and *Penicillium* in the rhizosphere leading to the control of *Sclerotium rolfsii.*

B. Foliar application of fertilizers and agrochemicals

Translocation of photosynthate from leaves to roots takes place as a part of the normal metabolic activity in plants. Therefore, organic substances, including plant protection chemicals (fungicides, insecticides), growth regulators and plant nutrients applied to foliage / leaves get absorbed into the leaf tissue and further get translocated to roots along with photosynthates. Many workers have reported that foliar application with various chemicals cause marked alterations in the number and kind / qualities of microorganisms in the rhizosphere of several cereals and leguminous crop plants. Thus, such an approach can be used as a new tool in the biological control of root diseases, stimulation of activity of nitrogen-fixing bacteria and other beneficial microorganisms in the soil.

C. Seed treatment with bio inoculants

Bio inoculants such as *Azotobacter, Beijerinckia, Azospirillum, Rhizobium* or P -solubilizing microorganisms (eg. *Bacillus, polymyxa, Azotobacter croococcum, Aspergillus niger, Penicillium digitatum* etc.) When applied to the seed /soil helps in the establishment of beneficial microorganisms in the rhizosphere region which will further benefit in plant growth, encourage inhibition of plant pathogenic organisms in the root vicinity and enrich the soil with added microbial bio-mass.

Associative and Antagonistic Activities in the Rhizosphere

In natural environments (eg. Soil, Air, Water etc.) a number of relationships exist between individual microbes, microbial species and between individual cells. The composition of microflora of any habitat (soil / rhizosphere) is governed by the biological equilibrium created by the associations and interactions of all individuals found in the community. In soil and rhizosphere region, many microorganisms live in close proximity and their interactions with each other may be associative or antagonistic.

A. Associative interactions / activities in rhizosphere: The dependence of one microorganism upon another for extra-cellular products (eg. amino acids and growth promoting substances) can be regarded as an associative activity / effect in rhizosphere. There is an increase in the exudation of amino acids, organic acids and monosaccharide by plant roots in the presence of microorganisms. Gibberellins and gibberellin- like substances are known to be produced by bacterial genera viz *Azotobacter, Arthrobacter, Pseudomonas,* and *Agrobacterium* which are commonly found in the rhizosphere. Microorganisms also influence root hair development, mucilage secretion and lateral root development. Fungi inhabiting the root surface facilitate the absorption of nutrient by the roots.

Mycorrhiza is one of the best known associative / symbiotic interactions which exist between the roots of higher plants and fungi. This mycorrhizal association has been found to improve plant growth through better uptake of phosphorus and zinc from soil, suppression of root pathogenic fungi and nematodes. Another example is association between the bacterium *Rhizobium* and roots of legumes and *Azospirillum* with cereal crops (wheat, rye, bajara, maize etc).

B. Antagonistic interactions / activities in rhizosphere: The biochemical qualities of root exudates and the presence of antagonistic microorganisms, plays important role in encouraging or inhibiting the soil borne plant pathogens in the rhizosphere region. Several mutualistic, communalistic, competitive and antagonistic interactions exist in the rhizosphere. The number and qualities of antagonistic microorganisms in the rhizosphere could be increased through artificial means such as fertilizer application, organic amendments, foliar spraying of chemicals etc.

Antagonistic microorganisms in the rhizosphere play an important role in controlling some of the soil borne plant pathogens. Scientists discovered the bacterial strain *Pseudomonas fluorescens* and the fluorescent pigments of this species in biological control of root pathogens. Strains of *P. fluorescence* are collectively called as **"Fluorescent Pseudomonads"**. They produce variety of biologically active compounds such as plant growth substances, cyanides, antibiotics and iron chelating substances called "Siderophores". These are low molecular weight, extra cellular, iron-binding agents produced by pseudomonads in response to low iron stress or when Fe^3 is in short supply. Thus, iron stress triggers the formation of iron-binding ligands called siderophores. Siderophores contains the pigments Pyovirdin (Fluorescent) and Pyocyanin (non-Fluorescent) having iron chelating properties. Another pigment "Pseudobactin" is a fluorescent chelator of iron which is known to promote plant growth and inhibition of pathogenic bacteria in the rhizosphere. An antibiotic called "Pyrrolnitrin" reduces damping-off disease in cotton caused by *Rhizoctonia solani.* Several species of *Bacillus* are known to cause mycolysis in the rhizosphere. *eg. Fusarium oxysporum* hyphae are known to undergo lysis in soil due to these bacterial metabolites.

The successful antagonists among fungi are *Trichoderma* sp (*T. viride* and *T. harzianum, T. hamatum*) and *Gliocladium virens* which parasitize, lyse or kill the phytopathogenic fungi in the soil. Antifungal and antibacterial actinobacteria in the rhizosphere play an important role in controlling pathogenic fungi and bacteria, for example *Micromonospora globosa* is a potent antagonist of *Fusarium udum* causing wilt of pigeon pea. Amoebae are also known to play an antagonistic role in controlling soil fungi, eg. control of take-all disease of wheat caused by *Gaumannomyces graminis* through the use of Myxamoebae. There can also occur antagonisms between two fungi producing metabolite and interfering the growth of the other fungus as in case of *Peniophora* antagonizing *Heterobasidium*.

Rhizosphere in Relation to Plant Pathogens

Plant root exudates influence pathogenic fungi, bacteria and nematodes in various ways. The effect may be in the form of attraction of fungal zoospores, or bacterial cells towards the roots; stimulation of germination of dormant spores and hatching of cysts of nematodes. Root exudates may contain inhibitory substances preventing the establishment of pathogens. The balance between the rhizosphere microflora and plant pathogens and soil microflora and plant pathogens is important in host-pathogenic relationship. In this context, the biochemical qualities of root exudates and the presence of antagonistic micro-organisms plays an important role in the proliferation and survival of root infecting pathogens in soil either through soil fungistasis, inhibition or antibiosis of pathogens in the rhizosphere.

Some of the most common interactions between plant roots and plant pathogenic microorganisms in the rhizosphere are discussed herewith.

A. Zoospore attraction: Amino acids, organic acids and sugars in the root exudates stimulate the movement and attraction of zoospores towards root of the plants. For example attraction of zoospores has been reported in *Phytophthora citrophthora* (Citrus roots), *P. parasitica* (tobacco roots) and *Pythium aphanidermatum* (pea root).

B. Spore germination: The spores or conidia of many pathogenic fungi such as *Rhizoctonia, Fusarium, Sclerotium, Pythium, Phytophthora* etc. have been stimulated to germinate by the root exudates of susceptible cultivars of the host plants. There are some reports on the selective stimulation of *Fusarium, Pseudomonas* and root infecting nematodes in the rhizosphere region of the respective susceptible hosts. This stimulus to germination is especially important to those plant pathogens which are not vigorous competitors and remain in resting stage due to shortage of nutrients or fungistasis. As a rule, germination and subsequent hyphal development are promoted by non host species and also by both susceptible and resistant cultivars of the host plants. The quantity and quality of microorganisms present in the rhizosphere of disease resistant crop varieties are significantly different from those of susceptible varieties.

C. Changes in morphology and physiology of host plant: Changes in the physiology and morphology of host plant influence the rhizosphere microflora through root exudations. Hence, significant changes in the rhizosphere microflora of diseased plants were reported which are attributed to the nature and severity of the disease. Systemic virus diseases cause marked changes in the plant morphology and physiology to drastically alter the rhizosphere microflora.

D. Increase in antagonists activity: Root exudates provide a food base for the growth of antagonistic organisms which plays an important role in controlling / suppressing some of the soil borne plant pathogens. Generally, rhizosphere of the resistant plant varieties harbour more number of *Streptomyces* and *Trichoderma* than that of susceptible varieties. For example in the rhizosphere of pigeon pea varieties resistant to *Fusarium udum,* the population of *Streptomyces* was found more which inhibited the growth of the pathogen. High density of *Trichoderma viride* in the rhizosphere of Tomato varieties resistant to *Verticillium* wilt has been reported with its ability to reduce the severity of wilt in susceptible plants.

E. Inhibition of pathogen: Root exudates containing toxic substances such as glycosides and hydrocyanic acid may inhibit the growth of pathogens in the rhizosphere. It has been reported that root exudates from resistant varieties of Flax (eg. Bison) excrete a glucoside which on hydrolysis produces hydrocyanic acid that inhibits *Fusarium oxysporum*, the flax root pathogen. Exudates of resistant pea reduce the germination of spores of *Fusarium oxysporum*.

In this light, the rhizosphere may be considered as a microbiological buffer zone in which the microflora serves to protect the plants against the attack of the pathogens.

F. Attraction of bacteria and nematodes: Root exudates attracts phytopathogenic bacteria and fungi in the rhizosphere for example *Agrobacterium tumefaciens* have been reported to be attracted to the roots of the host plants like peas, maize, onion, tobacco, tomato and cucumber.

Host root exudates also influence phytopathogenic nematodes in two ways: (i) though stimulation of egg-hatching process and (ii) attraction of larvae towards plant roots.

19

Spermosphere and Phyllosphere

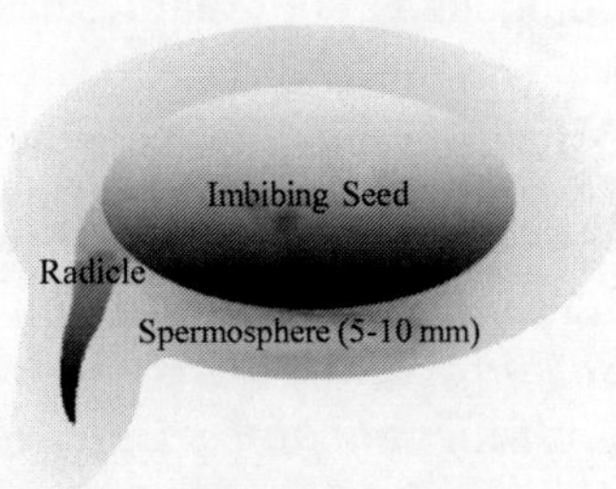

The spermosphere represents a short-lived, rapidly changing, and microbiologically dynamic zone of soil surrounding a germinating seed. It is analogous to the rhizosphere, being established largely by the carbon compounds released into the soil once the seed begins to hydrate. These seed exudations drive the microbial activities that take place in the spermosphere, many of which can have long-lasting impacts on plant growth and development as well as on plant health.

The germinating seed and surrounding soil represents a rich habitat for microbial development and interaction. The main energy source for microorganisms in this habitat is the carbon released by the seed into the surrounding soil. The habitat known as the spermosphere is the zone of microbial interaction around the seed.

The spermosphere was first mentioned in a study of seedling pathogens of forage grasses, in which Slykhuis (1947) noted that "the development of *Fusarium culmorum* in the environment in the immediate vicinity of germinating seeds was different from that in the surrounding soil." He defined this region as the "spermatosphere," being aware of the microbiological uniqueness of this region and speculating that the spermatosphere was of particular importance in regulating the activities of seed and seedling pathogens.

Seed Exudates

Seed swelling that results from the rapid influx of water leads to considerable internal hydrostatic pressures. This leads not only to rupture of the seed coat, but also to leakage of internal substances from the seed, because of the temporary structural damage to tonoplast and plasma lemma membranes when in a dried state. This rapid leakage of cellular and vacuolar constituents is referred to as seed exudation.

The imbibitional processes largely drive the exudation of molecules from the seed and into the surrounding soil and serve to define the types of molecules that are released. Since seed exudation is greatest during the first few minutes after imbibition commences, the types of molecules released during this period are generally low-molecular-weight compounds, which were preformed during seed development and maturation. Additionally, larger molecules such as peptides and proteins can be released because of the hydrostatic pressures in the inside of the seed. A wide range of sugars, amino acids, organic acids, phenolic compounds, and volatiles has been identified.

Microflora in the Spermospere

Species of *Fusarium* and *Pythium* were the dominant spermoplane/spermosphere fungi recovered from turnip seeds germinated for 72 h in soil. *Rhizoctonia solani* and species of *Penicillium*, *Trichoderma, Gliocladium, Cylindrocarpon, Cephalosporium, Cunninghamella, Mucor,* and *Helicocephalum* were recovered sporadically and at low frequency. These same fungi were also isolated in roughly the same proportions from tomato, onion, cabbage, bean, mustard, and melon spermospheres.

Among the bacteria colonizing barley seeds during the early stages of germination are species of *Acinetobacter, Bacillus, Burkholderia, Pantoea,* and *Pseudomonas*, whereas cottonseeds were colonized by species of *Xanthobacter, Enterobacter, Microbacterium, Paracoccus, Curtobacterium, Micrococcus, Agrobacterium, Paenibacillus,* and unidentified coryneform bacteria.

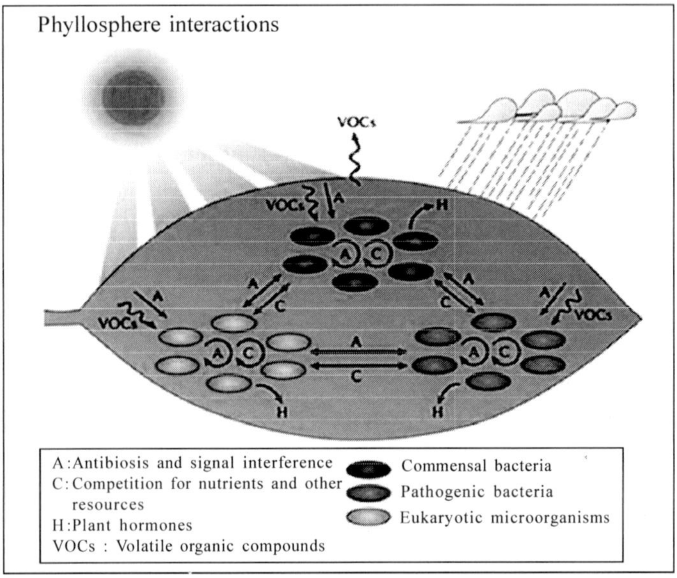

Phyllosphere

The surface of leaf is called phylloplane and the zone on leaf inhabited by microorganisms is called phyllosphere. The plant leaves are exposed to dust and air current that result in the establishment of typical flora on the surface of leaf.cuticles, wax, leaf appendages and leaf exudates help in anchorage of microorganisms on the surface of leaf. These microorganisms may die, survive, multiplies on leaf depending on extent or influence of leaf exudates on microorganisms.Leaf exudates contains aminoacids, glucose, fructose and sucrose that provides nutrition for growth of phyllospheric microorganisms. Phyllospheric microorganisms also promote growth of plant by various ways.

Phyllospheric microorganisms

The surface of leaf contains large number of bacteria, fungi, actinobacteria and algae as given below;

Bacteria: *Beijerinckia, Erwinia, Azotobacter, pseudomonas, Pseudobacterium, Phytomonas, Serratia etc*

Blue green algae: *Anabaena, Calothrix, Nostoc, Scytonema, Tolypothrix*

Fungi: *Alternaria, Cladosphorium, Cercospora, Penicillium, Saccharomyces*

Actinobacteria: *Streptomyces*

Factors Influence Phyllospheric Microorganisms

Various environmental factors and other conditions affect growth of microorganisms on the surface of leaf. They include.

1. Leaf exudates

- Leaf exudates contains many microbial growth factors such as amino acids, glucose, fructorse, sucrose etc that facilitates growth of phyllospheric microorganisms.
- Moisture released during transpiration provides water for growth of microorganisms on leaf.

2. Position of leaf

- Position of leaf also affects phyllospheric microorganisms.
- The surface of leaf which is directly exposed to sunlight contains relatively low microorganisms than the other surface of leaf which is shaded. It is due to antimicrobial effect of UV light.

- Furthermore, surface of leaf which is directly exposed to sunlight is relatively dry and moisture is less available for the microbial growth.

3. Leaf appendages

- Leaf appendages such as trough and veins affect growth of phylospheric microorganisms.
- Trough represents shallow depression on the surface of leaf. Microorganisms attach easily and grow in trough. Therefore the number of microorganisms is relatively higher on depressed area on the leaf.
- Veins present in leaf form depression in one surface and hence influence number of microorganisms.

4. Stomatal cavities

- In stomata water is in the form of drops that provides moisture for growth of phyllospheric microorganisms. Therefore number of microorganisms is relatively high in and around the stomatal cavities.

5. Age and types of plants

- With age of plant, rate of secretion of exudates is altered that affect growth of phyllospheric microorganisms, number of stomata and hence rate of transpiration is different in different plant species that affects growth of phyllospheric microorganisms.

6. Environmental pollutants

- Environmental pollutants deposited on the surface of leaf in the form of dust influence growth of phyllospheric microorganisms.
- For examples; cement and fertilizers are anti-microbial and they decrease number of phyllospheric microorganisms.

7. Insecticides, Antibiotics, herbicides etc.

- Insecticides, antibiotics, herbicides and other chemical sprayed on leaf on agricultural crops are antimicrobial and decreases the number of phyllospheric microorganisms.

8. Phytoalexin

- Phytoalexins are the chemicals produced by plants leaf that inhibits phyllospheric microorganisms as well as other microorganisms.

- Nature of phytoalexin differ from plant to plant.
- Phytoalexin is produced by plant in response to certain stimuli called elicitor.
- Elicitors may be biotic or abiotic factors.
- Biotic elictors include various chemicals produce by pathogenic and other phyllospheric microorganisms.
- Abiotic elicitors includes UV light, cold, tissue damage etc. When phytoalexin is produced by leaf, number of phyllospheric microorganisms decreases.

Role of Phyllospheric Microorganisms

- Phyllospheric microorganisms are important for plant growth. They promote growth of plants by various ways as given below:
- Some phyllospheric microorganisms such as Cyanobacteria and nitrogen fixing bacteria such as *Azotobacter* fix atmospheric Nitrogen and provide it to plant growth.
- Phyllospheric microorganisms produce various plant growth hormones such as indole acetic acid (IAA), which is used by plant for their growth.
- Phyllospheric microorganisms provide stimulus for production of phytoalexin by plants. Phytoalexin is the defensive chemical produced by plants that kill pathogenic as well as other microbes.
- Phyllospheric microorganisms decompose leaf and help in formation of humus after falling of leaves from plant.
- Some phyllospheric organisms have antagonistic effects against fungal pathogen and hence protect the plants from fungal diseases.
- The phyllospheric microorganisms colonize the surface of leaf forming a layer. Therefore these organisms compete with pathogenic microorganisms for habitat and nutrients.
- Some phyllospheric microorganisms degrade wax and cuticle of leaf and damage it.

Effect of plants on phyllospheric microorganisms

Surface of leaf provides natural habitat for growth and multiplication of phyllospheric microorganisms. Various leaf appendages as well as wax, cuticles and leaf exudates help in attachment of phyllospheric microorganisms to the surface of leaf.Aminoacids, glucose, sucrose etc present in leaf exudates provides nutrition for growth of phyllospheric microorganisms. Similarly water released during transpiration provides moisture for growth.Plants usually support growth and activities of phyllospheric microorganisms. However sometimes plant produce phytoalexins that kills many phyllospheric microorganisms.

Methylotrophs

Methylotrophs are a diverse group of microorganisms that can use reduced one-carbon compounds, such as methanol or methane, as the carbon source for their growth; and multi-carbon compounds that contain no carbon bonds, such as dimethyl ether and dimethylamine. This group of microorganisms also includes those capable of assimilating reduced one-carbon compounds by way of carbon dioxide using the ribulose bisphosphate pathway.

These organisms should not be confused with methanogens which on the contrary produce methane as a byproduct from various one-carbon compounds such as carbon dioxide. Some methylotrophs can degrade the greenhouse gas methane, and in this case they are called methanotrophs. The methanotroph *Methylococcus capsulatus* is used to degrade methane and other pollutants. The abundance, purity, and low price of methanol compared to commonly used sugars make methylotrophs competent organisms for production of amino acids, vitamins, recombinant proteins, single-cell proteins, co-enzymes and cytochromes.

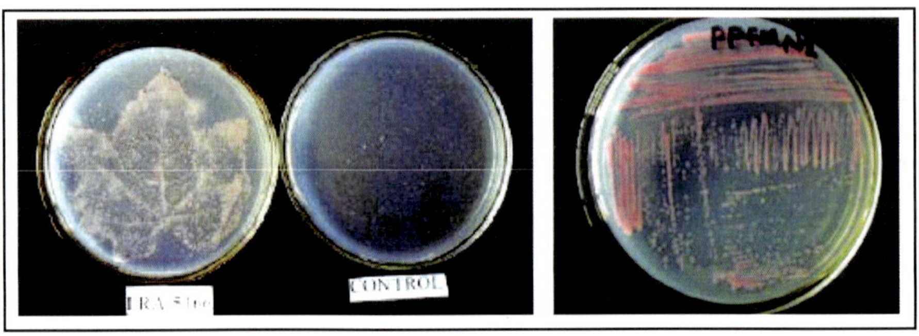

Methylobacteria isolation from phyllosphere by leaf-imprint method

Methylotrophic organisms are able to grow at the expense of one or more of the compounds typically used by methanophiles, but cannot use methane. Methylotrophs are a diverse group, including both Gram-negative and Gram positive genera. None of them make resting structures like exospores and cyst and none of them have the complex intracellular membrane systems that characterize methanotrophs growing on methane.

There are two sub groups:

1. Obligate methylotrophs.
2. Facultative methylotrophs.

Obligate Methylotrophs

A single obligate methylotroph (Methylophilus) is known. It is Gram-negative, polarly flagellated rod capable of rapid growth with methanol. Some strains can also utilize formaldehyde or methylamines. Carbon is assimilated *via* the ribulose mono phosphate pathway.

Facultative Methylotrophs

It is relatively widely distributed trait among heterotrophic bacteria. It may also be common among chemoautotrophs: several thiobacilli and nitrifying bacteria can drive CO_2 assimilation via the Calvin-Benson cycle by formate oxidation.

Examples of methylotrophs:

Genus *Methanosarcina*

Methylococcus capsulatus

Bacteria of the genus *Methylobacterium* are well-studied examples of facultative methylotrophs. These bacteria are classified as proteobacteria and are capable of grow on methanol and methylamine as well as on a variety of C_2, C_3, and C_4 compounds. Bacteria often live in environments containing diverse substrates. One of such habitat is the phyllosphere, where facultative methylotrophic bacteria are found to be ubiquitous and abundant. These methylotrophic bacteria belong to the genus *Methylobacterium* and are known to metabolize methanol but also a limited number of alternative carbon substrates, such as organic acids and alcohols. Because of their distinctive pink pigmentation, they are sometimes referred to as PPFMs (pink-pigmented facultative methylotrophs).

PPFMs are aerobic, Gram-negative bacteria and, although they are able to grow on a wide range of multicarbon substrates, they are characterized by the capability to grow on one-carbon compounds such as formate, formaldehyde, and methanol as their sole carbon and energy source. They are easily isolated on a methanol based mineral medium. Recently published data suggest that the degree of the plant–*Methylobacterium* association varies from strong, or symbiotic to loose, or epiphytic, a range that also includes the intermediate endophytic association. The use of leaf impressions on a methanol-agar medium has demonstrated the presence of pink-pigmented facultative methylotrophs in over 70 plant species. These PPFMs are especially abundant on leaves of field-grown crops averaged about 10^6 cfu of PPFMs per leaflet, and typically >80% of the viable bacteria recovered from leaves were PPFMs. The occurrence of soil methylotrophs is probably related to the abundance of plant lignin and pectin in soils; these polymers are major potential sources of methanol.

Plant leaf surfaces release diverse carbon sources, mainly sugars and organic acids, at low amounts (μm range), and these sources are heterogeneously located and result of leaching through the cuticle. In addition to these substrates, volatile carbon substrates, particularly the plant cell wall metabolism byproduct methanol, are released via the stomata. Methanol emission peaks in the morning, when the stomata first open. There is evidence that methanol is consumed by *Methylobacterium* and contributes to the epiphytic fitness of the organism. However, in addition to the peak of methanol emission in the morning, *Methylobacterium* should adapt its metabolism to use additional carbon sources during the rest of the day when methanol emission is low or during the night when stomata are closed and methanol is consequently no longer available.

A number of reports suggest that *Methylobacterium* strains are more than passive passengers on plant leaves. For instance, they have been shown in some cases to stimulate seed germination and plant development, possibly by production of phytohormones and it has been reported that one strain produces the cytokinin zeatin while others have been reported to produce indole acetic acid. In addition, *Methylobacterium* strains have been suggested to contribute to the flavor of strawberries and have been localized as endosymbionts within cells of the buds of Scotch pine (*Pinus sylvestris*).

20

Endophytic Microorganisms

Microbes that live inside plant tissues without causing disease are termed as endophytes. Endophytes have been found in all parts of plant *viz*., seeds, roots, stems, leaves, fruits etc. Within these organs, there is evidence of endophytes living in apoplastic intercellular space within parenchyma tissues and xylem vessel apoplast. Endophytic bacteria have been isolated from both monocotyledonous and dicotyledonous plants, ranging from woody tree species to herbaceous crop plants. Further, a single plant may be a host for several different bacterial species. They are consistently present in the root, stem, leaf, fruit and tuber tissues of a wide range of agricultural, horticultural and forest species. All plants in nature harbor a diverse community of endophytic bacteria which can positively affect host plant growth.

The potential of endophytic bacteria to fix nitrogen, promote plant growth and suppress plant diseases has renewed interest in such associations. It is advantageous to have the endophytic plant growth promoting organisms safely inside the plant, which could supply substantial amounts of nitrogen and other nutrients, growth hormones and other beneficial effects to the crop plants without the interference of the rhizosphere soil environment. Furthermore, plant growth promotion is often greater when it is induced by endophytes rather than by rhizosphere bacteria.

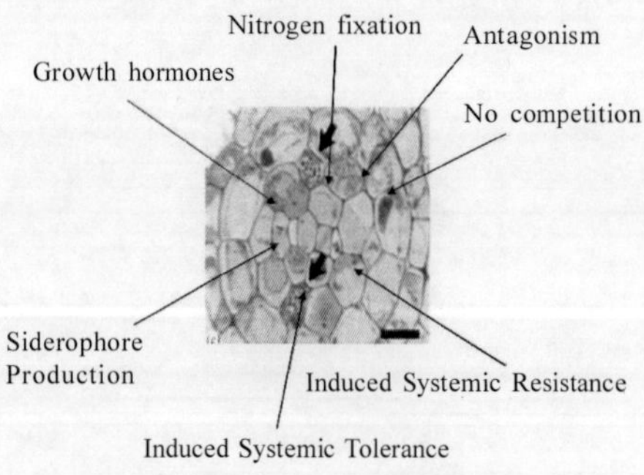

Beneficial attributes of endyphytic bacteria

Colonization

Endophytic microbes prior to their entry into plant tissues first colonize the plant roots by chemotaxis and electrotaxis and through accidental encounter. However, microbial colonization on root surface involves three steps: 1. bacterial adsorption onto the root surface, 2. anchoring on the root surface, and 3. colonization of the root surface with some cells entering into the root. However, endophytes easily enter at the point of breakage of the epidermis, colonize at the cortex, and then spread into vascular tissue across the endodermis. Sometimes, endophytes release the cell wall degrading enzymes to get enter in to the root. They subsequently colonized the root intercellular spaces, aerenchyma and cortical cells. Few endophytes are penetrating the stele to enter the vascular tissue. The xylem vessels in leaves and stems were also colonized.

Bacterial Genera with Endophytic Nature

Acetobacter, Acidovorax, Achromobacter, Acinetobacter, Actinomyces, Agrobacterium, Alcaligenes, Arthrobacter, Azoarcus, Azorhizobium, Azospirillum, Bacillus, Bordetella, Burkholderia, Capnocytophaga, Cellulomonas, Citrobacter, Chryseobacterium, Clavibacter, Comamonas, Corynebacterium, Curtobacterium, Deleya, Enterobacter, Escherichia, Erwinia, Flavobacterium, Herbaspirillum, Kingella, Klebsiella, Lactobacillus, Leuconostoc, Methylobacterium, Micrococcus, Moraxella,Pantoea, Pasteurella, Photobacterium, Phyllobacterium, Providencia, Pseudomonas, Psychrobacter, Rahnella, Rhizobium, Rhodococcus, Serratia, Shewanella, Sphingomonas, Staphylococcus, Stenotrophomonas, Variovorax, Vibrio, Xanthomonas, Yersinia.

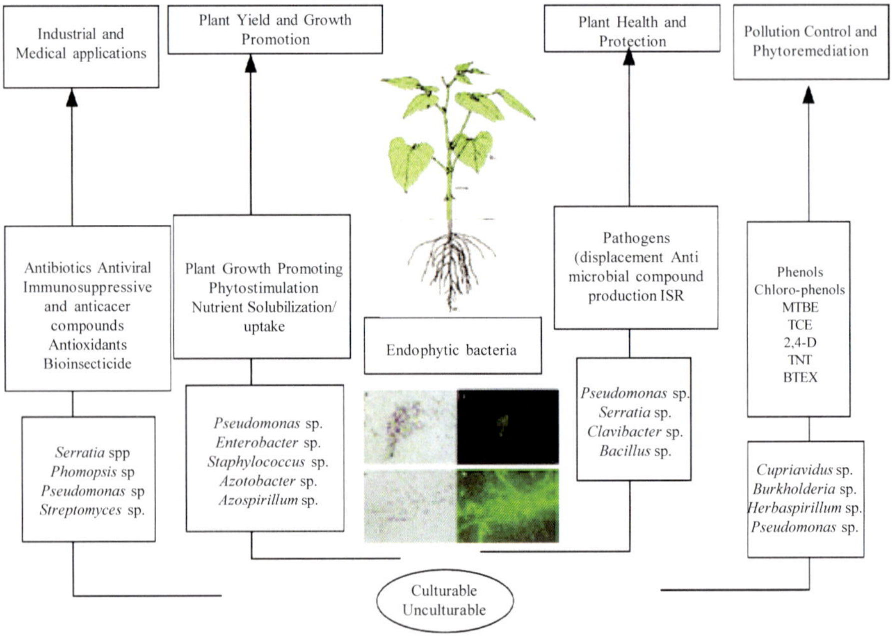

Endophytes on plant growth promotion

Being in close interaction with the plant, endophytic bacteria can directly or indirectly promote the plant growth by nutrients mobilization, synthesis of plant growth hormones and also help to combat with other pathogens and pests. Endophytic bacteria benefit plants using various traits, including synthesis of plant growth regulators, osmoprotectants, exopolysaccharides, antifungal metabolites and the modulation of plant physiobiochemical constituents. Endophytes benefit plants through nitrogen fixation (BNF), phytohormone production, siderophore production, ACC-deaminase activity, nutrient solubilization/availability, vitamin production, and suppression of phytopathogens. The endophytic bacteria assimilate atmospheric nitrogen and convert it into ammonia, transferring this molecule to the plant metabolism. The phytohormones *viz*., auxins (IAA), gibberellins (GAs), cytokinins, abscisic acid (ABA), and ethylene are the best known plant hormones secreted by many endophytes.

The endophytes release organic acids like 2-ketogluconic acid, gluconic acid, etc. which lower the pH and ultimately solubilize the insoluble P and make them available to crops. Several endophytes produce siderophores, a low molecular weight iron-chelating compounds that holds Fe3+ and helps in Fe acquisition. Different kinds of siderophores such as ornibactin, ferrichrome, and desferrioxamine are produced by different species of endophytes. Production of 1-aminocyclopropane-1-carboxylate (ACC) deaminase is one of the important characteristics of plant growth-promoting endophytes. ACC deaminase cleaves ACC, the immediate precursor of the plant hormone ethylene, to produce á-ketobutyrate and ammonia. This trait of endophytic bacterial isolates can be attributed to significant growth enhancement of crop plants in the problem soil (stress) environment.

Endophytes as Biocontrol Agents

Endophytic bacteria are able to penetrate and become systemically disseminated in the host plant, actively colonizing the apoplast, conducting vessels, and occasionally the intracellular spaces. This colonization presents an ecological niche, similar to that occupied by plant pathogens, and this endophytic bacteria can, therefore, act as biological control agents against pathogens. Numerous reports revealed that endophytic microorganisms control plant pathogens, insects and nematodes. Endophytic bacteria can antagonize soil-borne pathogens through various mechanisms such as competition, antibiosis, and/or parasitism. Bacterial endophytes prevent disease development through endophyte-mediated *de novo* synthesis of novel compounds and antifungal metabolites.

21

Plant Growth Promoting Rhizobacteria (PGPR)

Plant Growth Promoting Rhizobacteria (PGPR) are root-colonizing bacteria that form symbiotic relationships with many plants. The name comes from the Greek *rhiza*, meaning root. Though parasitic varieties of rhizobacteria exist, the term usually refers to bacteria that form a relationship beneficial for both parties (mutualism). They are an important group of microorganisms used in biofertilizer. Biofertilization accounts for about 65% of the nitrogen supply to crops worldwide. The term PGPRs was first used by W. Kloepper in the late 1970s and has become commonly used in scientific literatures. PGPRs have different relationships with different species of host plants. The two major classes of relationships are rhizospheric and endophytic. Rhizospheric relationships consist of the PGPRs that colonize the surface of the root, or superficial intercellular spaces of the host plant, often forming root nodules. Endophytic relationships involve the PGPRs residing and growing within the host plant in the apoplastic space.Some common examples of PGPR genera exhibiting plant growth promoting activity are: *Pseudomonas, Azospirillum, Azotobacter, Bacillus, Burkholdaria, Enterobacter, Rhizobium, Erwinia, Mycobacterium, Mesorhizobium, Flavobacterium*, etc.An ideal PGPR should possess high rhizosphere competence, enhance plant growth capabilities, have a broad spectrum of action, be safe for the environment, be compatible with other rhizobacteria, and be tolerant to heat, UV radiation, and oxidizing agent.

Role of PGPR in Plant Growth Improvement

PGPR plays an important role in enhancing plant growth through a wide variety of mechanisms. The mode of action of PGPR that promotes plant growth includes i. abiotic stress tolerance in plants; ii.nutrient fixation for easy uptake by plant; iii.plant growth regulators; iv.the production of siderophores; v. the production of volatile organic compounds; and vi.the production of protection enzymes such as chitinase, glucanase, and ACC-deaminase for the prevention of plant diseases. However, the mode of action of different PGPR varies depending on the type of host plants.

i. Abiotic stress tolerance in plants

Abiotic stresses are considered to be the main sources of agricultural yield reduction. The intensity of abiotic stress varies depending on the type of soils and plant factors. PGPR could be attenuated by the toxic effect of chemical pollution on plants due to the binding ability of the bacteria with the ions from the soil, thereby decreasing the availability of chemical in the soil. Another major effect of PGPR on plants under abiotic stress conditions is the improvement of leaf water status, especially under salinity and drought stress. The ability of plants in utilizing water for growth depends on their stomatal apertures. The stomata on the plant leaf functions to balance the water content in leaf and water uptake by the roots. The stomatal conductance (water vapour exiting through the stomata leaf) of plant leaf was higher in PGPR inoculated plants than non-PGPR inoculated ones under drought conditions. It is evident that PGPR-inoculated plants tend to improve the water-use efficiency of plants.

ii. Nutrient fixation for easy uptake by plant

PGPR has the ability to increase the availability of nutrient concentration in the rhizosphere byfixing nutrients, thus preventing them from leaching out.

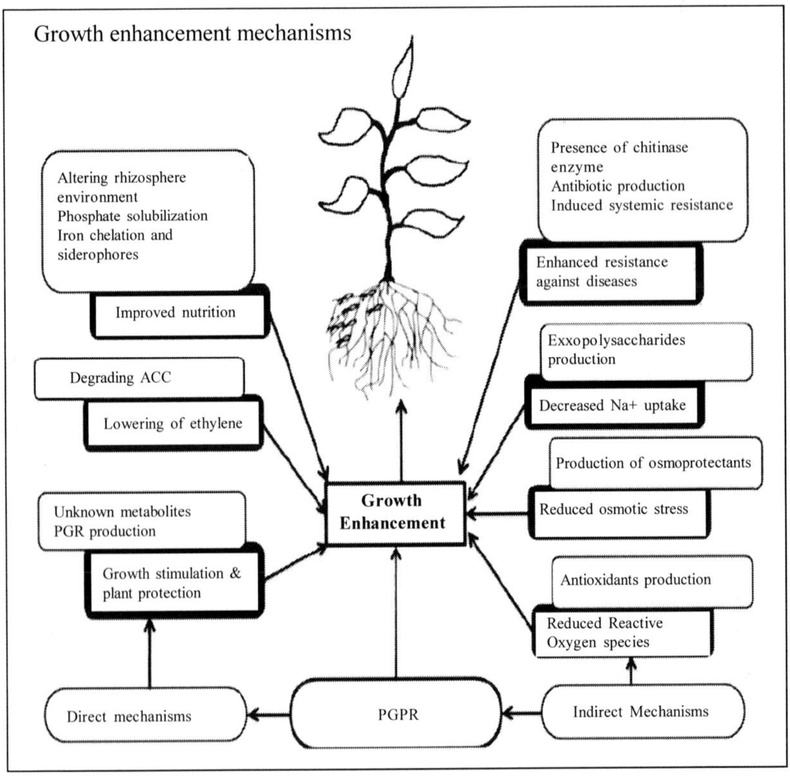

Nitrogen fixation

Nitrogen fixation is one of the most beneficial processes performed by rhizobacteria. Nitrogen is a vital nutrient to plants and gaseous nitrogen (N_2) is not available to them due to the high energy required to break the triple bonds between the two atoms. Rhizobacteria, through nitrogen fixation, are able to convert gaseous nitrogen (N_2) to ammonia (NH_3) making it an available nutrient to the host plant which can support and enhance plant growth. The host plant provides the bacteria with amino acids so they do not need to assimilate ammonia. The amino acids are then shuttled back to the plant with newly fixed nitrogen.

Phosphate Solubilization

These bacteria secrete different types of organic acids (e.g., carboxylic acid) thus lowering the pH in the rhizosphere and consequently release the bound forms of phosphate like $Ca_3(PO_4)_2$ in the calcareous soils. Utilization of these microorganisms as environment-friendly biofertilizer helps to reduce the use of expensive phosphatic fertilizers.They also help to increase the availability of accumulated phosphate (by solubilization), increase the efficiency of biological nitrogen fixation and render availability of Fe, Zn, etc., through production of plant growth promoting substances.

iii. Production of plant growth regulators

One of the prominent modes of action for plant growth promotion by PGPR is phytostimulation or plant growth regulation. These microorganisms have the ability to produce the growth hormones (phytohormones) such as indole-3-acetic acid (IAA), indole-3-butyric acid (IBA),gibberellic acid (GA3), cytokinins, zeatin,abscisic acid (ABA) and ethylene. Phytohormones are organic substances found in extremely low amounts that exert influence on the biochemical, physiological, and morphological processes in plants.

iv. Production of siderophores

PGPR are reported to secrete some extracellular metabolites called siderophores. The significance of siderophores produced by certain genera of PGPR in plant growth promotion is well reported by many workers. Siderophores are commonly referred to as microbial Fe-chelating low molecular weight compounds. The presence of siderophore producing PGPR in rhizosphere increases the rate of Fe^{3+} supply to plants and therefore enhance the plant growth and productivity of crops.Further, this compound after chelating Fe^{3+} makes the soil Fe^{3+}deficient for other soil microbes and consequently inhibits the activity of competitive microbes.

v. Production of volatile organic compound

Volatile organic compounds (VOCs) produced by plant growth promoting rhizobacteria (PGPR) are heavily involved in improving plant growth and induce systemic resistance (ISR) towards pathogens. Several bacterial species, from diverse genera including *Bacillus*, *Pseudomonas*,*Serratia*, *Arthrobacter*, and *Stenotrophomonas* produce VOCs that influence plant growth. Acetoin and 2,3-butanediol synthesized by *Bacillus* are the best known of these compounds and are responsible for significant improvements in plant growth.

vi. As biocontrol agents

PGPR promote plant growth through the control of phytopathogenic agents, primarily for the production of metabolites contributing to the antibiosis and antifungal properties used as defense systems. The mechanism would involve the production of hydrolytic enzymes, of which two examples are chitinase and glucanase. Major fungal cell wall components are made up of chitin and beta-glucan, thus chitinases and beta-glucanases producing bacteria would inhibit fungal growth.PGPR may protect plants against pathogens by direct antagonistic interactions between the biocontrol agent and the pathogen, as well as by induction of host resistance.

PGPR that indirectly enhance plant growth via suppression of phytopathogens by a variety of mechanisms. These include, a. The ability to produce siderophores that chelate iron, making it unavailable to pathogens, b. The capacity to synthesize anti-fungal metabolites such as antibiotics or hydrogen cyanide which suppress the growth of fungal pathogens c. The ability to successfully compete with pathogens for nutrients or specific niches on the root d. the ability to induce systemic resistance.

22

Soil Microbes and Their Interactions

Soil is the largest terrestrial ecosystem where a wide variety of relationships exists between different types of soil organisms. The associations existing between different soil microorganisms, whether of a symbiotic or antagonistic nature, influence the activities of microorganisms in the soil. Microflora composition of any habitat is governed by the biological equilibrium created by the associations and interactions of all individuals found in the community. In soil, many microorganisms live in close proximity and interact among them-selves in a different ways. Some of the interactions or associations are mutually beneficial, or mutually detrimental or neutral.

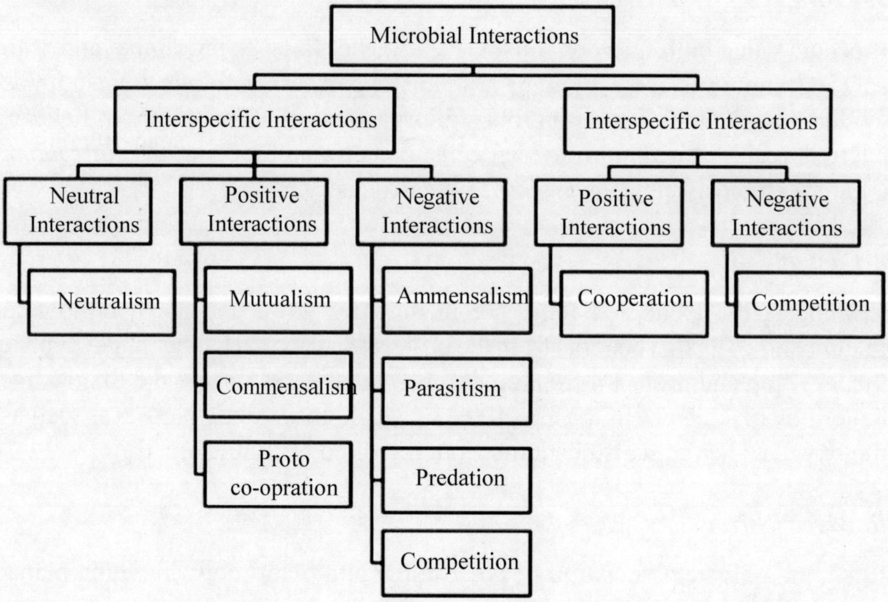

A. Neutralism

It is a type of neutral association, in which two dissimilar organisms inhabiting the same environment without impacting other microorganisms and behave as entirely independent. Each could utilize different nutrients without producing metabolic end products that are inhibitory.

Eg: Some bacteria may exhibit neutralism as it has been reported that some species of *Lactobacillus* and *Streptococcus* can coexist without affecting each other positively or negatively.

B. Symbiosis / Mutualism

It is an obligatory or highly specific interaction between two populations in which both of them benefit from each other. It usually requires close physical connection in which both partners can act, as if they are one, when in separate their metabolic activities are different.Based on the partner selection it is classified as obligate symbiosis and facultative symbiosis. Based on the purpose of interaction also it is classified.

1. Obligate symbiosis

It occurs when both microorganism live together in close proximity, and both species cannot survive without its mutualistic partner.Symbiotic association is evident in soil among several groups of organisms algae and fungi in lichens, bacteria residing within protozoan cells, bacteria and roots in the *Rhizobium* legume symbiosis, fungi and roots in mycorrhizae.

i. Lichens

In lichens, the algae and fungi are in intimate physical and physiological relationship. Thealga benefits from the protection afforded by the fungal hyphae that envelop and protect it from environmental stresses. While the fungi gains benefit by making use of the CO_2 fixed by its photosynthetic partner as well as the oxygen.The fungi also benefits from the fixed N_2 of algae.

ii. Mycorrhizae

It is a mutualistic association between fungi and plant roots, in which plants provide fungus with carbohydrates and offer it protection. In turn, the fungus increases the surface area of plant roots for absorbing water, nitrogenous compounds, phosphorus, and other inorganic nutrients (e.g., phosphate)from the surrounding soil and delivers them to the plant which improves plant growth and health.

iii. Symbiotic N_2 Fixation

The® nitrogen-fixing bacteria provide the plants with nitrogenous compounds, while in return the plants provide the nitrogen-fixing bacteria with carbohydrates. This mutualistic association improves plant growth and health, and it has different types which include *Rhizobium* spp. with root nodules of legume plants and *Frankia* which is an actinobacteria (nodule-forming filamentous bacteria) with the roots of *Alnus* and *Casuarina* trees which are "nonlegumes"

2. Facultative symbiosis

It occurs when one of the two partners can survive without its mutualistic partner by itself in some conditions. Membership in this association is not usually specific and one organism can be replaced by the other. It is a loose association.

C. Protocoperation

Synergism (protocooperation) is a relationship that occurs between two or more populations at which both or all of them benefit. In this relationship microbial populations perform a function which may not be performed individually or produce a new product that neither each population can produce alone.This relationship is different from mutualism because as it is not an obligatory interaction, none of the species depend on the relationship for existence, as each member can live and produce its own food individually.

Types of protocooperation

There are different types of protocooperation relationship that could be found in the terrestrial ecosystem which is considered very useful in agriculture:

- Nutritional protocooperation

It is the most popular relationship between terrestrial populations at which the populations exchange nutrients between each other. Such cooperation is also called syntrophism.

Nutritional proto co-operation has been demonstrated in cultures. Eg: In a medium deficient in nicotinic acid and biotin, neither *Proteus vulgaris* nor *Bacillus polymyxa* will multiply as the former bacterium requires nicotinic acid and the latter needs biotin. In mixed culture, in the same medium however both grown since the partner bacterium synthesizes the missing vitamins.

Protocooperation also occurs between higher plants growing in the soil and bacteria or fungi living in the rhizosphere. Neither each of them is dependent on this association, since bacteria and fungi get benefits from the exudates of plant roots and interact with each other to form the essential nutrients necessary for plant's growth such as decomposed organic materials, production of phytohormones, minerals, water, vitamins and amino acids which in turn improve soil fertility as well as the plant health and growth.

- ***Metabolism of toxic end products***: In this type of association one organism establishes its association byeliminating toxic substances from the habitation versus obtaining carbon products made by the otherassociate partner.
- ***Production of derivative enzymes***: *Arthrobacter* and *Streptomyces* (soil flora) produce enzymes whichcollectively degrade diazinon which is an organophosphate pesticide (useful in the degradation ofxenobiotics or recalcitrant compounds).

D. Commensalisms

It is a relationship at which one population benefits, while the other population is unaffected (neither harmed nor benefited). It is a very common relationship between different microbial populations.This occurs commonly in soil with respect to degradation of complex molecules like cellulose and lignin. One partner can attack a substrate not available to the second organism, but the decomposition resultsin the formation of products utilized by the second. The one which offer benefit is called commensal. eg: Many fungi able to degrade cellulose and yield glucose and organic acids. This can serve as a nutrient source for many bacteria and fungi, which are noncellulolytic

Negative / harmful / deleterious interactions

Detrimental effects of one species on its neighbours are quite common in soil, and they are demonstrated by the decreases in abundance or metabolic activities of the susceptible organisms. It consists of different relationships between different populations either two or more, at which one population at least is harmed while the other is either harmed, benefited, or not affected.

This include

a) Competition

b) Amensalism

c) Parasitism

d) Predation

a. Competition

It is a relation that occurs between different populations in the soil which use the same limiting resources that are insufficient to support all the individuals. The rivalry for limiting nutrients or other common needs. These resources include raw materials important for life such as water, light, nutrients, oxygen, and space for occupying or any other resources, which is essential for survival and reproduction.In this relation, the best adapted microbial species will predominate or infact, eliminate other species whichare dependent upon the same limited nutrient substances. Also, organisms which have the capability to grow faster are considered good competitors.

Resource competition It occurs when the growth rates of both populations are limited by the same resource and one population has the ability to diminish the availability of that resource for the other populations.

Interference competition It occurs between two populations in which one of them damages the other population's habitat either physically or chemically and excludes it from the habitation. This relationship is also called direct competition or active competition.

The chlamydospores of *Fusarium*, oospores of *Aphanomyces*, as well as conidia of *Verticillium dahliae* need exogenous nutrients to germinate in soil, while some other fungi, bacteria that inhabit the soil have the ability to deplete these important nutrients that are required for spore germination and thereby delay its germination which results in decrease in the population of these plant pathogens inthe soil

b. Amensalism (Antagonism)

It is the most common negative relationship in nature at which one microbial population suppresses or adversely influences the growth or the activities of the other population in the same environment by producing inhibitory substances either directly or indirectly. The release of products by one species is toxic to its neighbours. The population that produces the inhibitors is not affected by them and therefore gains the antagonistic edge.These inhibitors may be antibiotics, toxins, organic acids, alcohols, or other allelochemicals, lytic enzymes, as well as harmful gases like methane, ethylene, HCN, nitrite, or sulfides or other volatile sulfur compounds. The population that adversely affects the other is called antagonistic species,and it constantly has great practical importance. Antagonism is a type of amensalism.

Types of Antagonism

There are diverse types of antagonism according to the nature of substances that is used in the antagonism.

- ***Antagonism by Antibiosis***: This process is called antibiosis in which the antibiotics or other allelochemical metabolites are produced by one organism to inhibit another organism. Antibiotics is common among *Streptomyces* isolates, but numerous strains of *Micromonespora* and *Nocardia* are also active. *Bacillus* and *Pseudomonas* species are producing antibiotic substances against plant pathogenic fungi. Other than antibiotic some inhibitory compounds are secreted by several soil microorganisms.
- Antimicrobial compounds against fungi are present in the soil, which inhibit the germination of fungal spores. This phenomenon is termed as **fungistasis**.
- Cyanide is produced by certain fungi in concentrations toxic to other microorganisms.
- Algae elaborate fatty acids which exhibit and marked antibacterial activity.
- *Bacillus thuringiensis* releasetoxin to lepidopteran insect pests.

c. Parasitism

It is a relationship between two dissimilar organisms that is called host-parasite relationship in which one of them (parasite) lives in or on the other organism (host).The parasite lives in close contact with the host and forms metabolic association with the host and feeds on their cells, tissues, or fluids in which the parasite is profited, while the host is adversely affected. Sometimes the relation between the host and parasite could be diverged from parasitic relationship to a pathogenic relationship.

This relationship is widely spread in soil communities and characterized by its long period of contact and the specialization between parasite and host. Also, parasite is usually smaller than the host (in most cases). This relationship has two sides, one is useful while the other harmful. If the parasitism is accomplished on bacteria that are considered pathogenic to plants, it is considered as a useful relationship for plant growth and health. While if the parasitism is accomplished on bacteria that are considered profitable to plants, it is considered as a harmful relationship for plant growth and health.

Types of Parasitism

Parasitism could be classified according to its nature of parasitism and its infection type.

Types of Parasitism according to Parasitism Patterns

- *Obligate parasitism*: Occurs when the parasite cannot live without its host.
- *Facultative parasitism*: Occurs when the parasite can survive by itself without its host cells in some conditions.

Types of Parasitism according to Infection Form

- *Ectoparasitism*: The parasite remains outside the host cells.
- *Endoparasitism*: The parasite penetrates the host cells.

Examples of Parasitism

(a) Viruses which attack bacteria (bacteriophages), fungi, algae, or plants are strict endoparasites(intracellular parasites) as they obligate parasite and cannot be cultivated on the media as free-living forms.

(b) Chytrid fungi parasitize on algae as well as other fungi by penetration into the host.

(c) *Daptobacter* spp. penetrates and degrades the cytoplasm of several genera of *Chromatiaceae*. It grows and propagates in the cytoplasm of the host

d. Predation

Predation is one of the most dramatic interrelationships among the microorganisms in nature, at which predator organism directly attacks a prey organism and feeds on it. This relationship has short duration, at which predators may or may not kill their prey prior to feeding on them, but the normal result is generally absorption of the prey's tissue through ingestion and subsequently the death of prey. Prey may be larger or smaller than predator.

The most numerous predators of bacteria are protozoans, which by feeding on the billions of bacteria, undisputedly affect their populations. Protozoans are a key factor in limiting the size of bacterialp opulations.

Bdellovibrio bacteriovorus is a predatory bacterium, which penetrates the cell wall and multiplies between the wall and the plasma membrane, which causes lysis of the prey and releases its progeny. It attacks and consumes different bacterial strains, including *Escherichia coli* and *Aquaspirillum serpens*, *Salmonella typhimurium*, and *Helicobacter pylori*.

Trichoderma spp. could be considered as the most essential biological control agent in the soil because it utilizes different mechanisms to fight various disease-causing agents including parasitism, which directly attack the pathogen especially fungi; competition, as it has the ability to colonize the soil or to compete for nutrients which is causing relegation of pathogen from plant rhizosphere; antagonism (antibiosis), as it is able to produce secondary metabolites,which have a lethal or depressant effect on the plant pathogens.

23

Industrially Important Microorganisms

Industrial microbiology is an important area of applied microbiology. In a broader sense, industrial microbiology is concerned with all aspects of business that relate to microbiology. It refers to the use of microorganisms in commercial enterprise. Cheap raw materials are converted to valuable products through the metabolism of microbes. Microbes for this purpose could be exploited in different ways.

Various commercial products of economic value made by microbes are :

(i) medicines i.e. pharmaceuticals, including antibiotics, steroids, human protein, vaccines, and vitamins,

(ii) organic acids,

(iii) amino acids,

(vi) enzymes,

(v) alcohols,

(vi) organic solvents

(vii) synthetic fuels.

(viii) recovery of metals from ores through bioleaching,

(ix) recovery of petrol, and

(x) single cell protein production.

The term fermentation in industrial microbiology is used in a wider sense to include any chemical transformation of organic compounds carried out by using microbes and their enzymes. Production methods in industrial microbiology bring together the raw materials (substrates), microorganisms (specific strains or microbial enzymes) and a controlled favourable environment (created in a fermentor) to produce the desired substance. The essence of an industrial process is to combine the right organism, an inexpensive substrate, and the proper environment to produce high yields of a desired product.

Microbes possess a wealth of metabolic equipment that brings about diverse chemical transformations. This characteristic of microbes could be exploited in obtaining some valuable products of daily use. The cheap raw materials available in nature as a waste may be converted into useful commercial products by the activity of microbes. Microbes thus serve a dual purpose, firstly, they are good agents of disposal of these wastes, and secondly the resultant end products of their breakdown are useful commercial products.

Characteristics of Industrially Important Microbes

- The microbes used must be contamination free. In case, if it is bacteria, it should be phage free.
- Grow in simple media: minimal nutritional requirements. Also, preferably without growth factors (i.e., pre-formed vitamins, nucleotides, and acids).
- The organism should be genetically and physiologically stable. Hence, they must resist random mutations.
- The organism should also accept a certain degree of genetic manipulation to enable the creation of strains with more acceptable properties.
- The microbes should grow vigorously and rapidly.
- The growth period must be shorter, not more than 24 - 36 hours.
- They should lead to a single desired product in a short time possible.
- The product should not contain unwanted materials and toxins.
- The microbes should produce their products extracellularly to reduce the cost price.
- They should produce the products of interest in high yield.
- If possible, the microbes used should be nonpathogenic.
- The organism should not be too highly demanding of oxygen.
- The microbes should be conserved for an extended period and become viable after thawing process.
- The microbes need to be reasonably resistant to predators such as *Bdellovibrio* spp or bacteriophages.

Microbes in Industrial Products

These microbes play a crucial role in the fermentation process to obtain a number of products. The common products obtained by fermentation process through industrial processes are fermented beverages, malted cereals, broths, fruit juices, antibiotics, etc.

Beverages

Yeasts are the widely used microorganisms for the production of beverages like beer, brandy, rum, wine, whiskey, etc. Yeasts are single-celled, eukaryotic, microorganisms of the Kingdom Fungi. In these industrial process, the species of yeasts, *Saccharomyces cerevisiae*, generally called as the Brewer's Yeasts are used for fermenting fruit juices and malted cereals to produce ethanol. Once after the fermentation, these beverages are distilled to produce both alcoholic and non-alcoholic beverages including whiskey, brandy, rum, etc.

Yeast - *Saccharomyces cerevisiae*

Bacteria - *Zymomonas mobilis*

Organic acids

Microbes are also used for the industrial production of certain organic acids. Citric acid was the first discovered organic acids from microbial fermentation of lemon – a citrus fruits. Organic acids are also produced directly from glucose. *Aspergillus niger*, *Acetobacter acute* and *Lactobacillus* are few examples of microbes used for the industrial production of organic acids.

Microbes and substrates used in organic acid production

Type of organic acid	Substrate	Microorganism
Acetic acid	Sugarcane juice	*Acetobacter* sp.
Citric acid	Beet molasses	*Aspergillus niger*
Citric acid	Alkane	*Saccharomyces cerevisiae*
Gluconic acid	Glucose	*Aspergillus niger*
Lactic acid	Whey	*Lactobacillus delbrueckii*
Lactic acid	Glucose	*Lactobacillus* sp.

Enzymes

Enzymes are naturally occurring, biological catalysts that are mainly used to control certain biochemical reactions in the living system. Enzymes have a wide range of applications in the production of both medical and non-medical field. Apart from the plants and animals, enzymes are also obtained from certain microbes and are referred to as the microbial enzymes. Microorganisms are majorly used for the production of industrial enzymes through the safe gene transfer methods. The first industrially produced microbial enzymes were obtained from the fungal amylase in the year 1896 and were used to cure indigestion and several other digestive disorders.

Microbes and substrates used in enzymes production

Type of enzyme	Substrate	Microorganism
Amylase	Starch	*Saccharomyces diastaticus*
Protease	Proteins	*Bacillus* sp.
Lipase	Lipids	*Aspergillus* sp.
Pectinase	Pectin	*Aspergillus niger*, *Penicillium* sp.
Cellulase	Cellulose	*Cellulomonas* sp.
Xylanase	Xylan	*Streptomyces* sp.
Chitinase	Chitin	*Actinomyces*, *Streptomyces* sp.
Sucrase	Sucrose	*Streptomyces* sp.
Phosphatase	Tricalcium phosphate	*Phosphobacterium*, *Pseudomonas* sp.

Antibiotics

Antibiotics are chemical substances produced by certain microbes which functions either by killing or retarding the growth of harmful microbes without affecting the host cells. Penicillin was the first antibiotic to be discovered by Alexander Fleming in the year 1928 from the fungus *Penicillium notatum*. There are many other antibiotics produced by microorganisms, including Streptomycin, and other antibiotics used to treat a number of bacterial infections.

Antibiotic	Microbial source	Antibiotic spectrum	Inhibitory action
Penicillins	*Penicillium* spp.	Gram- positive bacteria	Cell wall synthesis
Steptomycin	*Streptomyces griseus*	Gram- positive bacteria and Gram-negative bacteria	Protein synthesis
Tetracycline	*Streptomyces aureofaciens*	Broad spectrum	Protein synthesis
Bacitracin	*Bacillus subtilis*	Gram- positive bacteria	Cell wall synthesis
Chloram-phenicol	*S. Venezuelae*	Broad spectrum	Protein synthesis
Cycloheximide	*S. griseus*	Fungi	Protein synthesis
Nystatin	*S.nouresii*	Fungi	Cell membrane

Vitamins

Vitamins are organic compounds which are capable of performing many life-sustaining functions inside our body. They are essential micronutrients which are required in small quantities for the body's metabolism. As our body cannot synthesize these vitamins, they need to be supplied through the diet. Apart from plants and animals sources, microbes are also capable of synthesizing the vitamins.

There are few groups of microbes living in the digestive tracts of both humans and other animals which are collectively called as the gut microbiota. These microbes are involved in synthesizing vitamin K. Other examples of microbial vitamins include ascorbic acid, beta-carotene, biotin, ergosterol, folic acid, vitamin b12, Thiamine, pantothenic acid, riboflavin, and pyridoxine.

Vitamin C - *Corynebacterium* sp.

Biotin - *Serratia marcescens, Bacillus sphaericus*

Riboflavin - *Eremothecium ashbyii, Ashbya gossypii, Bacillus* sp.

Vitamin B12 - *Propionibacterium shermanii, Pseudomonas denitrificans*

Vitamin E - *Euglena gracilis*

Food

Following are some of the well known food materials and the respective microbes involved in the production.

Product	Organism
Cultured butter milk	*Streptococcus* and *Leuconostoc*
Cultured sour cream	*Streptococcus* and *Leuconostoc*
Bulgarian milk	*Lactobacillus bulgaricus*
Acidophilus milk	*L. acidophilus*
yogurt	*S. thermophilus, L. bulgaricus*
Kefir	*S. lactis, L. bulgaricus*
Cottage cheese	*S. lactis*
Cheddar, swiss chesse	*S. faecalis*
Cambert- soft cheese	*Penicilliumcaneberti, Brevibacterium*
Roquefort- semi soft	*Penicillium roqueforti*

Single Cell Protein

Single cell protein (SCP) typically refers to sources of mixed protein extracted from pure or mixed cultures of algae, yeasts, fungi or bacteria (grown on agricultural wastes) used as a substitute for protein-rich foods, in human and animal feeds. They have average protein contents of 44% to 60%.

Microbes employed include

Yeasts: *Saccharomyces cerevisiae, Candida utilis, Torulopsis*and *Geotrichum candidum*

Fungi: *Aspergillus oryzae, Sclerotium rolfsii, Polyporus and Trichoderma*

Bacteria: *Rhodopseudomonas capsulate*

Algae: *Chlorella* and *Spirulina.*

24

Silage Production

Silage is preserved pasture. Fresh forage crops such as maize, grasses, legumes, wheat and lucerne can be preserved by ensiling. Silage is the green material produced by controlled fermentation of the green fodder crop retaining the high moisture content, which contain 20-40% dry matter.

Silage can be prepared from plants having thick stems and are generally not very suitable for hay making like sorghum, maize. Weeds can also be utilized along with main fodder crops for silage making. Silage making kills majority of weed seeds. It is highly palatable. The organic acids produced in the silage are similar to those normally produced in the digestive tract of the ruminants and therefore are used in the same manner.

Principles of Silage Making

Anaerobic condition is first and foremost requirement for silage making, as it allows lactic acid bacteria to grow, which converts sugars into lactic acid, a strong organic acid. As pH declines, the degrading actions of plant enzymes and undesirable bacteria are inhibited and at pH 3-4, most degrading enzymes are inhibited and the growth of lactic acid bacteria is also inhibited.

Sugars + Oxygen ⟶ Carbon dioxide + Water + Heat

This reaction is prevented and nutrient loss is reduced and in simple words the principle of silage is the same as that in making pickles.

Selection of crops for silage making

It is easier to ensile forages that have:

- High level of fermentable sugar.
- Low level of protein.
- Low buffering capacity.
- Should have about 35 per cent dry matter at the time of ensiling.

Pasture grasses

Elephant grass (Napiergrass), Guinea grass, Rhodes grass, Sudan grass, Setaria, Ruzi grass, etc.

- Pasture legumes: Stylosanthes guianensis
- Fodder tree: Leucaena.
- Straws: rice straw, wheat straw, soybean chaff, peanut hulls, etc.
- Corn, Sorghum.
- Farm by-products.

Site for construction of silo (silage storage structure)

The place where we intended to construct silo should be elevated from other places. The place must be free from water clogging. Walls of silo pit must be leak proof. It should at least six feet away from form animal shed. If it is possible construct the silo south side of the animal shed. The stages of crops used for silage is as follows.

Pasture grasses: From early heading to heading stage.

Pasture legumes: From budding to early flowering stage.

Corn: Yellow ripening stage.

Sorghum: Dough stage.

Procedure for silage making

Construct a surface/ trench silo. One cubic meter space silo can store 500- 600 kg of green fodder. First, chopping of forage to short length (1-3 cm), so that the packing density is kept higher. Nearly 65-70 % moisture level in green fodder is ideal for silage making. High moisture content leads to poor fermentation. Spread the chopped fodder to make a bed of 2 feet height in Silo-Pit. Compact forage as tightly as possible with help of tractor or hand roller or any heavy object. Sprinkle Molasses solution over it at 2-3% of the material weight. If the moisture content of the raw material is high, molasses addition may have no effect.

Repeat steps at least 5 to 10 times till the Silo- pit is completely filled up. Seal the pit airtight with plastic cover. Plastic covers should be pressed with heavy objects like bricks or tire. Maintain sealing for 45 days. Once silo-pit will be open, it should be finished within 45-60 days to avoid fungal contamination.

Fermentation Process of Silage

First stage (Respiratory stage)

The packed raw materials are still respiring immediately after chopped and consumes oxygen. The temperature will rise to about 32°C around 4 days after packing.

Second stage (Early fermentation)

Production of acetic acid, formic acid and other organic acids as a result of the growth of facultative aerobic bacteria such as Enterobacteria. The silage pH slowly changes from about 6.0 to 4.0.

Third stage (Lactic acid fermentation)

Lactic acid fermentation begins by lactic acid bacteria witch are strictly anaerobic about 3 days after packing chopped materials and acetic acid production declines.

Fourth stage (Continuation of Lactic acid fermentation)

Lactic acid production continues for about 2 weeks. The temperature goes down slowly to about the normal atmospheric temperature and pH is maintained at 4.0.

Fifth stage (Stabilization phase)

Due to the presence of lactic acid, further degradation is inhibited, as bacterial and fungal growths are checked. The lactic acid fermentation completes in about 20 days, and the silage product is finished.

Characteristics of good quality silage

- The final product silage must be bright, light green yellow or green brown in colour
- It should have lactic acid odour with no butyric acid and ammonia odour.
- The texture must be firm with soften material.
- Moisture content should be in the range of 65-70%.
- It should have lactic acid content of 3-14 % and butyric acid less than 0.2%.
- The pH must be around 4.0 to 4.2.

25

Biofertilizers and Their Types

Introduction

Biofertilizers are microbial preparations containing living or latent cells of efficient strains of nitrogen fixing, phosphate solubilizing and cellulose decomposing microorganisms intended for seed or soil application and designed to improve soil fertility and plant growth by increasing the number and biological activity of beneficial microorganisms in the soil.

The objects behind the application of biofertilizers /microbial inoculants to seed, soil is to increase the number and biological / metabolic activity of useful microorganisms that accelerate certain microbial processes to augment the extent of availability of nutrients in the available forms which can be easily assimilated by plants. The need for the use of biofertilizers has arisen primarily due to two reasons i.e. though chemical fertilizers increase soil fertility, crop productivity and production, but increased / intensive use of chemical fertilizers has caused serious concern of soil texture, soil fertility and other environmental problems.Use of biofertilizers is both economical as well as environment friendly. Therefore, an integrated approach of applying both chemical fertilizers and biofertilizers is the best way of integrated nutrient supply in agriculture.

Biofertilizers, thus include i) Symbiotic nitrogen fixers- *Rhizobium sp.* ii) Non-symbiotic, free living nitrogen fixers- *Azotobacter, Azospirillum* etc. iii) Blue Green Algae (BGA) inoculants- *Azolla-Anabaena,* iv) Phosphate solubilizing microorganisms (PSM)- *Bacillus, Pseudomonas, Penicillium, Aspergillus* etc. v) Mycorrhiza vi) Cellulolytic microorganisms

Role of Biofertilizers in soil fertility and Agriculture

Biofertilizers are known to play a number of vital roles in soil fertility, crop productivity and production in agriculture as they are eco friendly. Some of the important functions or roles of biofertilizers in agriculture are:

1. They supplement chemical fertilizers for meeting the integrated nutrient demand of the crops.

2. They can add 20-200 kg N/ha year *(eg. Rhizobium sp* 50-100 kg N/ ha year ; *Azospirillum , Azotobacter* : 20-40 kg N/ha /yr; *Azolla* : 40-80 kg N/ha; BGA :20-30 kg N/ha) under optimum soil conditions and thereby increases 15-25 percent of total crop yield.

3. They can at best minimize the use of chemical fertilizers not exceeding 40-50 kg N/ha under ideal agronomic and pest-free conditions.

4. Application of biofertilizers results in increased mineral and water uptake, root development, vegetative growth and nitrogen fixation.

5. Some biofertilizers *(eg, Rhizobium BGA, Azotobacter* sp) stimulate production of growth promoting substances like vitamin-B complex, Indole acetic acid (IAA) and Gibberellic acids etc.

6. Phosphate mobilizing or phosphorus solubilizing biofertilizers / microorganisms (bacteria, fungi, mycorrhiza etc.) converts insoluble soil phosphate into soluble forms by secreting several organic acids and under optimum conditions.They can solubilize / mobilize about 30-50 kg P_2O_5/ha due to which crop yield may increase by 10 to 20%.

7. Mycorrhiza (AM fungi) when used as biofertilizers enhance uptake of P, Zn, S and water, leading to uniform crop growth and increased yield and also enhance resistance to root diseases and improve hardiness of transplant shock.

8. They liberate growth promoting substances and vitamins and help to maintain soil fertility.

9. They act as antagonists and suppress the incidence of soil borne plant pathogens and thus, help in the bio-control of diseases.

10. Nitrogen fixing, phosphate mobilizing and cellulolytic microorganisms in bio-fertilizers enhance the availability of plant nutrients in the soil and thus, sustain the agricultural production and farming system.

11. They are cheaper, pollution free and renewable energy sources.

12. They improve physical properties of soil, soil tilth and soil health in general.

13. They improve soil fertility and soil productivity.

14. Blue green algae like *Nostoc, Anabaena, and Scytonema* are often employed in the reclamation of alkaline soils.

15. Bioinoculants containing cellulolytic and lignolytic microorganisms enhance the degradation/ decomposition of organic matter in soil.

16. BGA plays a vital role in the nitrogen economy of rice fields in tropical regions.

17. *Azotobacter* inoculants when applied to many non-leguminous crop plants, promote seed germination and initial vigour of plants by producing growth promoting substances.

18. *Azolla-Anabaena* grows profusely as a floating plant in the flooded rice fields and can fix 100-150 kg N/ha /year in approximately with 40-60 tones of biomass produced.

19. They play important role in the recycling of plant nutrients.

In soil the activities of nitrogen fixation, mobilisation of plant nutrients and degradation of ligno-cellulotic wastes are being carried out by a large number of micro-organisms. Artificially multiplied cultures of selected micro organisms (biofertilizers) augment the natural recycling of organic resources. There are different types of biofertilizers /microbial inoculants.

Classification of Biofertilizers

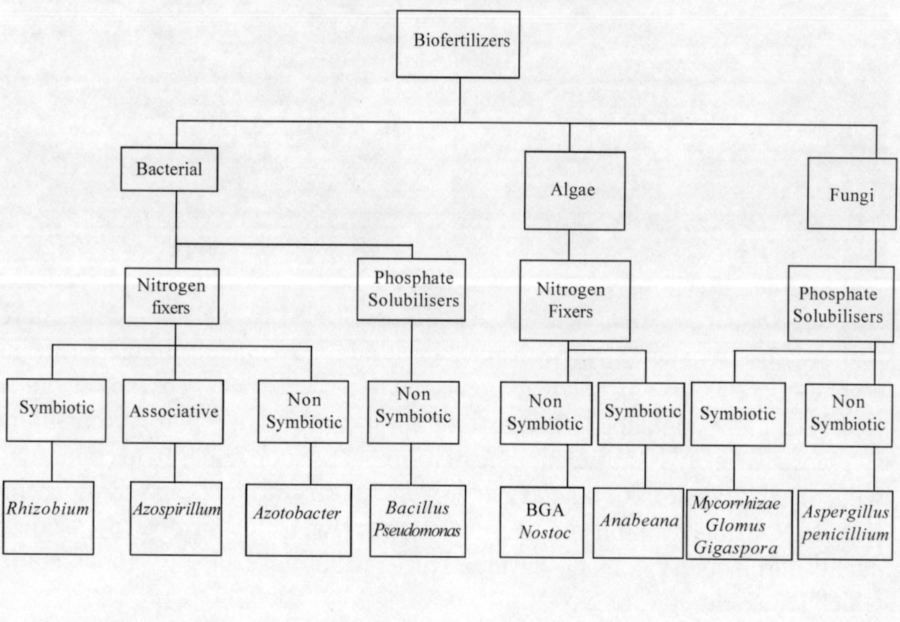

Nitrogen Fixing Microbes

Rhizobium

Legumes provide a protein - rich source of food, which is an essential part of the diet in many parts of the tropics. They play an important role in supplying those essential amino acids, particularly lysine that are not present in sufficient quantities in cereal crops. These leguminous plants in association with the soil bacterium, Rhizobium form root nodules and fix atmospheric N_2.It has been estimated that 40-250 kg N / ha / year is fixed by different legume crops by the microbial activities of *Rhizobium*.

Rhizobium infects the roots of leguminous plants. They are usually found in the soil and produce nodules after infecting the roots of the leguminous plants. As a result, nitrogen gas is fixed from the atmosphere. This nitrogen is made available to the plants that help in their growth and development. When the legume dies there will be a breakdown of nodules. As a result, *Rhizobium* is released back to the soil where it can infect a new host.

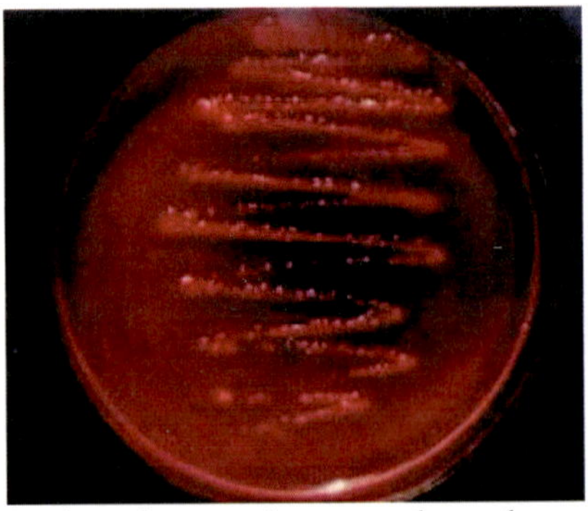

The rhizobia that form associations with the best known legumes have been characterized by nucleic acid sequence homologies as well as by the more traditional traits such as growth rates (fast and slow), acid production from carbohydrates, serology and legume associations. The success of these legume crops depends upon the efficient functioning of legume plant -*Rhizobium* symbiosis. The efficient functioning of nitrogen fixation under field conditions will contribute to improving the yield of legume crops. Yield can be influenced by several factors including the genotype of the plant, native rhizobial population, quality of rhizobial inoculant, and the environmental and ecological conditions in which legume crops are grown.

Cross Inoculation Groups of *Rhizobium* (CIG)

CIG refers the groups of leguminous plants that will develop effective nodules when inoculated with the rhizobia obtained from the nodules from any member of that legume group.

A. *Rhizobium*

		CIG	Host it can nodulate
1.	*Rhizobium leguminosarum*		
	bv. viceae	Pea	Peas,lenfils, vicia
	bv. phaseoli	Bean	*Phaseolus* spp
	bv. trifoli	Clover	*Trifolium* spp
2.	*R. meliloti*	Alfalfa	Alfalfa, clover, fenugreek
3.	*R. loti*	Lotus	Trifoli, lupine,
4.	*R. fredii*	Soybean	Soybean
5.	*R.* spp	Cowpea group	Vigna, Arachis, Cajanus, Dolichus, Sesbania, Acacia, Prosopis, green gram and blackgram
6.	*R.* sp	Chickpea group	Chickpea

B. Bradyrhizobium

B. japonicum	Soybean	Soybean
B. spp	Cowpea group	Cowpea group plants

C. ***Azorhizobium***- Stem nodulating – Nodulates *Sesbania rostrata.*

D. ***Photorhizobia*** - Nodulates*Aeschynomene* sp.

E. ***Sinorhizobium*** - fast growing soybean nodulator

To ensure the presence of adequate numbers of rhizobia in the root rhizosphere of legume crops, these are inoculated with superior rhizobial strains. The success or failure of rhizobial inoculants is largely governed by the combined interaction of various biotic and abiotic factors. Hence a large-scale screening should be carried out to identify ideal inoculant strain for different legume crops. The criterion for selection of efficient strain may vary for particular soil types like acidic, sodic-saline, nitrate rich or heavy metal contamination.

Azospirillum

The recognition in the mid seventies that strains of *Azospirillum* are abundantly associated with the roots of cereals and millets and help fixing substantial amounts of nitrogen led to the opening of the new line of work, the associative symbiosis. *Azospirillum* is considered to be an associative symbiotic bacterium because it lives in close association with the root system of plants. Almost all crop plants harbour this bacterium in their roots. These bacteria colonizing the roots not only remain on the root surface but also a sizable population penetrates into the root tissues and live in harmony with the plant. However they do not produce any visible nodules or outgrowth on the roots. In the absence of any plant,

Azospirillum lives in the soil just like any other microorganism saprophytically. When a suitable crop is raised they are attracted towards the root system. Making use of the nutrients exuded by the plants through their root system (root exudates) *Azospirillum* colonizes in the root region in large numbers and fixes substantial amounts of nitrogen.

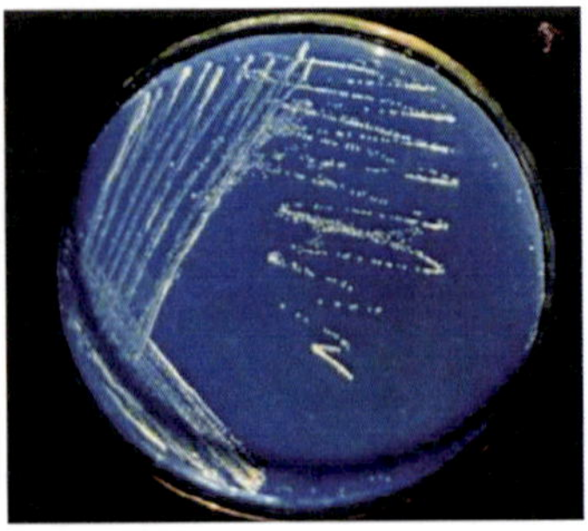

All azospirilla are nitrogen fixing bacteria with nitrogenase activity comparable to those of other nitrogen fixers. The bacteria are microaerophilic actively moving towards self-created gradients of oxygen concentration. Studies with inoculated plants in the field have shown that *Azospirillum* comprises 1-10% of the total bacterial numbers in the rhizosphere. It has been demonstrated that *Azospirillum* inoculated plants take up minerals (N, P and K) from solution at faster rates than controls and consequently inoculated plants in the field accumulate more dry matter, N, P and K in their stems and leaves. Under water stress conditions inoculated plant show improvement in the water content of the plants. Inoculated plants also recorded more highly branched and developed root system. From the above observations it is clear that *Azospirillum* inoculation benefits plant growth and increases yields of crops by improving root development, mineral uptake and plant water relationships. The effect of *Azospirillum* inoculation on the total yield increase of field grown crops generally ranged from 10 to 30 per cent. In addition to nitrogen fixation *Azospirillum* also produces growth promoting substances like IAA, GA, Kinetin, etc. and these phytohormones go a long way in enhancing the crop growth.

Azotobacter

These are free-living bacteria, which grow well on a nitrogen free medium. These bacteria utilize atmospheric nitrogen gas for their cell protein synthesis. This cell protein is then mineralized in soil after the death of *Azotobacter* cells thereby contributing towards the nitrogen availability of the crop plants. *Azotobacter* is Gram-negative bacteria, polymorphic i.e. they are of different sizes and shapes. Their size ranges from 2-10 x 1-2.5 µm, young cell possess peritrichous flagella and are used as locomotive organs. Old population of bacteria includes encapsulated forms and have enhanced resistant to heat, desiccation and adverse conditions. The cyst germinates under favourable conditions to give vegetative cells. They also produce polysaccharides. *Azotobacter* spp., are sensitive to acidic pH, high salts, and temperature above 35°C. Some of the important species of *Azotobacter* are *A. chroococcum, A. agilis, A.paspali* and *A. vinelandii* of which *A. chroococcum* is most commonly found in our soils.

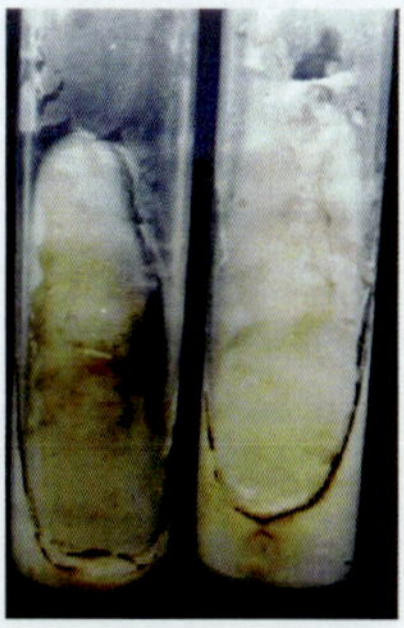

The species of *Azotobacter* are known to fix on an average 10 mg. of N/g of sugar in pure culture on a nitrogen free medium. A maximum of 30 mg. N fixed per gram of sugar was also reported. However, *Azotobacter* is a poor competitor for nutrients in soil. Most efficient strains of *Azotobacter* would need to oxidize about 1000 kg of organic matter for fixing 30 kg of N/ha. This does not sound realistic for our soils, which have very low active carbon status. Besides, soil is inhabitated by a large variety of other microbes, all of which compete for the active carbon.

In Indian soils, the population of *Azotobacter* is not more than 10 thousand to 1 lakh/g of soil. Other microorganisms present in soil mostly influence the Population of *Azotobacter*. There are some microorganisms, which stimulate the *Azotobacter* population in soil thereby increasing the nitrogen fixation by *Azotobacter*. On the other hand there are some microorganisms, which adversely affect the *Azotobacter* population, and hence nitrogen fixation process is hampered. For example *Cephalosporium* is most commonly found organisms in soil, which restricts the growth of *Azotobacter*. *Azotobacter* also produces some substances, which check the plant pathogens such as *Alternaria*, *Fusarium* and *Helminthosporium*. The problem faced with *Azotobacter* is that a substantial amount of energy is required by *Azotobacter* for nitrogen fixation. The possible source of energy is the soil organic matter. The energy generated during the decomposition of organic matter is used for nitrogen fixation. In soils where the level of organic matter is very low, *Azotobacter* runs short of energy source and ceases to fix nitrogen.

Gluconacetobacter

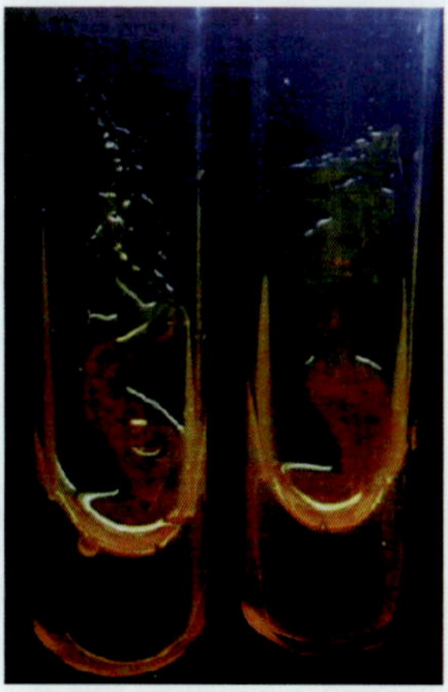

In recent years, explanation for large scale contribution of biological nitrogen fixation in crops like sugarcane and non-legumes has become much easier by description of obligate endophytic bacterium *Gluconacetobacter* diazotrophicus which get colonized within roots, stems and leaves of certain plant genotypes. In addition to *Gluconacetobacter diazotrophicus*, *Herbaspirillum seropedicae*, *Herbaspirillum rubrisubalbicans* and *Azoarcus* have also been reported as other endophytic bacteria. However physiological and nutritional studies with sugar rich crops viz., sugarcane, sweet potato, sweet sorghum and coffee plants revealed the importance of nitrogen fixing bacterium, *Gluconacetobacter diazotrophicus*. In addition to nitrogen fixation, these organisms also produce Indole-3-acetic acid, which serve as a growth promoter.

Salient features

- Acid tolerant nitrogen fixing bacterium associated with sugarcane.
- An endophytic organism colonizes the roots, stems, buds and leaves of sugarcane
- It is located in the apoplastic fluid of sugarcane stems and to a lesser extent in xylem vessels.
- Capable of growing and fixing nitrogen at 30% sucrose concentration and optimum sucrose concentration for better nitrogen fixation is 10%.
- Growth and Nitrogen fixation observed even at pH 3. 0.
- The optimum pH is 5.5.
- An average of 200 kg N ha^{-1} would be fixed by using 30 per cent of the total sugar produced by a sugarcane crop.
- IAA production is also noticed.
- Tolerant to higher level of combined nitrogen.
- Mode of entry is through mycorrhizae, mealy bug and mechanical injury of plant parts
- Exhibiting antibiotic resistance
- It is a member of α - Proteobacteria group
- In terms of 16s rDNA based phylogeny *Gluconacetobacter* clusters with the genera *Rhodopila*, *Acidomonas*, *Acidiphilium* and *Gluconobacter*
- In terms of n if H sequence, it is more related to *Rhizobium* sp. and *Azospirillum brasilense* than to other diazotrophs.

Phosphate Solubilizing Microorganisms (PSM)

Although phosphorus is one of the most important macronutrients of all living organisms, many processes of the P cycle are still obscure. Several P-forms might be taken up by the cell, but by far the greatest part is absorbed in the forms of HPO_4^{2-} or H_2PO_4. Therefore free inorganic phosphates in soil solution

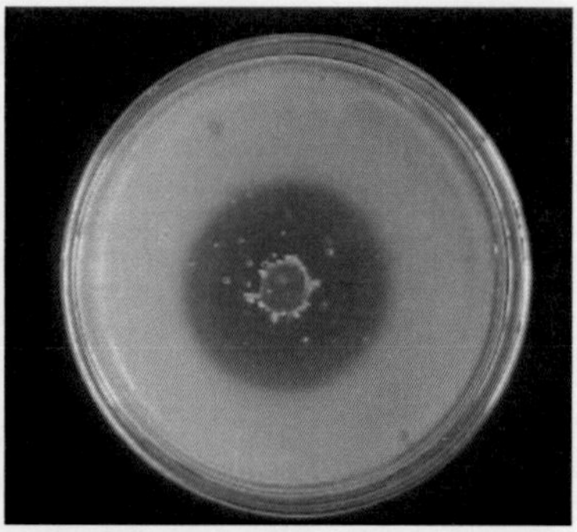

play a central role in the cycling of P. It has long been known that the microorganisms play an important role in the mobilization and immobilization of phosphorus. Phosphate solubilising microorganisms distributed well in several ecosystems, type and nature of microorganisms depend on the type of soil and the vegetation that comes on it. Various microorganisms such as bacteria, fungi, actinobacteria, cyanobacteria and yeast are known to solubilize insoluble inorganic as well as bound organic phosphates. A large fraction of soil microbial population can dissolve insoluble inorganic phosphates present in soils. In some soils this ability has been found to be present in as high as 8% of the total population.

The population of phosphate dissolving microorganisms is more in the rhizosphere (20-40% of the total population) compared to non-rhizosphere (10-15% of the total population). PSM along with acid production, produces the phosphatase enzyme which cause the solubilization of P. Besides these two mechanisms the production of chelating substances, H_2S, CO_2, mineral acids, siderophores are also involved in P solubilization by PSMs.

Phosphate Solubilizing Microbes

A large number of heterotrophic and autotrophic soil micro organisms known to have the capacity to solubilize inorganic phosphates through their metabolic activities directly (or) indirectly. The most efficient bacterial isolates were identified as *Bacillus megaterium* var *phosphaticum, Pseudomonas striata,* and *Bacillus polymyxa* and fungal isolates such as *Aspergillus awamori, Penicillium digitatum* and *Aspergillus niger.*

Azolla

Azolla is a green - floating water fern that floats in water and fixes atmospheric nitrogen in association with nitrogen fixing blue green alga *Anabaena azollae*. Rice growing areas in South East Asia and other third World countries have been evincing increased interest in the use of the symbiotic N_2 fixing water fern *Azolla* either as an alternate nitrogen source or as a supplement to commercial nitrogen fertilizers.

The eukaryotic water fern *Azolla* and the prokaryotic endosymbiont*Anabaena azollae* are actively involved in the symbiotic association and fix atmospheric nitrogen. The host plant *Azolla* provides the carbon source sucrose to the cyanobiont while the cyanobiont fix nitrogen and transfers to *Azolla* which in turn multiplies very rapidly. The cyanobiont is associated with dorsal lobes of the leaf from the onset of their development and is never in direct contact with the external environment and *A. azollae* is the only species found in symbiotic association with *Azolla*. The growth and development of the *Azolla* and the *Anabaena* are synchronous. *Azolla* shoot tips are supported above the water surface by the stem axes and a small colony of *Anabaena* filaments is associated with every apical merits. The heterocyst of the symbiont *Anabaena* is the site of nitrogen fixation, while vegetative cells are involved in photosynthesis. The total nitrogen requirement of the *Azolla* fronds is supplied by the cyanobiont.

The ammonia from N_2 fixation by the cyanobiont is released into the leaf cavity and absorbed by the frond's ammonia assimilating enzymes. The growth rate of *Azolla* is very rapid and it doubles its weight in a mere 2-3 days. Due to fast relative growth rate, the biomass production is heavy in a given time and hence it is exploited as biofertilizer for rice. *Azolla* has the potential of maintaining an exponential growth rate under optimum conditions.

Azolla is used as biofertilizer for wetland rice and it is known to contribute 40-60 kg N ha^{-1} per rice crop. Two methods of *Azolla* application have been recommended in India; as a green manure by incorporating in fields prior to rice planting and secondly by dualcropping with rice when the fern grows side by side with the main rice crop. The agronomic potential of *Azolla* is quite significant particularly for rice crop and it is widely used as biofertilizer for increasing rice yield. The important factor in using *Azolla* as a biofertilizer for rice crop is its quick decomposition in soil and efficient availability of its nitrogen to rice. In tropical rice soil, the applied *Azolla* mineralizes rapidly and its nitrogen is made available to the rice crop in very short period. Within 20-30 days *Azolla* can multiply and produce biomass ranging from 15-24 t ha^{-1}. The nitrogen content is ranging from 3-5% on dry weight basis which is relatively higher than other green manure plants. Similarly potassium accumulation is between 3-6% on dry weight basis.

Azolla can easily be grown along with rice in flood water and used as biofertilizer. *Azolla* can be incorporated easily when rice crop is in the field and it decomposes in the field rapidly and nitrogen is made available to rice plants. Since it produces spores i.e., microsporocarp and megasporocarps, the spores can withstand adverse conditions during summer and regenerate during cultivation cycle of rice crop. Therefore, once *Azolla* is introduced in rice it can sustain and act as a renewable biofertilizer. Besides nitrogen contribution to rice, *Azolla* also controls the weed flora in wet land rice, which is an economic advantage for rice growers, the residual effect of *Azolla* is useful for subsequent crops grown after rice. In place where water scarcity is acute, *Azolla* can putforth roots in rice field and act as mulch and conserve the moisture for 20-30 days. Besides its utilization as biofertilizer *Azolla* is also used as poultry and cattle feed since it has high protein and mineral content.

Cyanobacteria (Blue Green Algae)

Cyanobacteria otherwise called as Blue green algae (BGA) are oxygen evolving and nitrogen fixing prokaryotes, using sunlight as the sole energy source for the fixation of carbon and nitrogen. Many cyanobacteria also contribute greatly to the nitrogen economy of aquatic and terrestrial habitats through their ability to fix atmospheric nitrogen. Besides fixing atmospheric nitrogen, cyanobacteria play a major role in reducing soil erosion, increasing the soil organic matter content and in producing certain substances which enhance the growth of higher plants. Algal application resulted in a significant improvement in the aggregation status of the soil and brought down appreciably the pH, EC, CEC and increased the hydraulic conductivity of the soil.

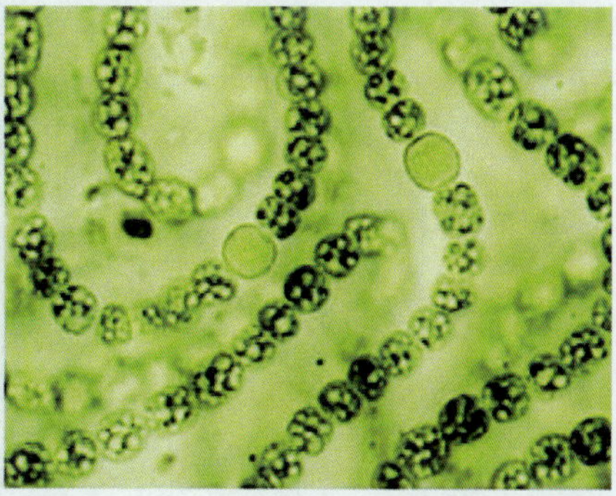

The rice field ecosystem provides the conductive environment for them with respect to light, water, high temperature and nutrient availability. Amongst the various soil factors affecting the distribution of cyanobacteria, pH is an important one which directly influence their distribution and abundance in soil. Cyanobacteria generally prefer neutral to slightly alkaline condition. Many exhibit salt tolerance due to the presence of extracellular polysaccharides. The salt tolerance of cyanobacteria has been exploited with some success in reclamation of saline and sodic soils. Application of cyanobacteria to saline soil enhances the soil organic carbon content due to the addition of enormous quantity of organic matter in the form of algal biomass and reduces the EC of saline soil. The salt tolerant cyanobacteria immobilize Na^+ in an organic form and thereby play a major role in saline soil.

The species of cyanobacteria which are known to fix atmospheric nitrogen are classified into three groups (i) Heterocystous – aerobic forms, (ii) Aerobic unicellular forms and (iii) Non heterocystous, filamentous, microaerophilic forms. The cyanobacterial cultures composite inoculum consisting of *Nostoc*, *Anabaena*, *Calothrix*, *Tolypothrix*, *Plectonema*, *Aphanothece*, *Gloeocapasa*, *Oscillatoria*, *Cylindrospermum*, *Aulosira* and *Scytonema* have been used for inoculation in rice. These N_2 fixing cultures have better adaptability in wetland rice.

Algal population appears to be highly susceptible to environmental changes and exhibit rapid qualitative and quantitative variation along the cultivation cycle. Development of dense algal blooms just after transplanting due to fertilizers or ploughing or both and a high light availability have been reported. *Anabaena* was the first species to develop two weeks after transplanting. N_2 fixing forms *viz., Scytonema*, *Aulosira*, *Nostoc*, *Aphanthece*, *Oscillatoria*, *Plectonema* were dominant during tillering phase of rice crop. During the cultivation cycle a decrease in light intensity and N level related to rice growth and an increase in pH value favoured cyanobacterial growth. *Gloeotrichia* and *Rivularia* are reported to occur during panicle initiation and flowering phase of rice.

Cyanobacteria is responsible for the spontaneous fertility of tropical rice field soils. Cyanobacterial inoculation was reported to be effective in different agro climatic conditions and soil types contributing 20 – 30 kg N/ha. Nitrogen fixed by the algae are released either through exudation or through microbial decomposition after the cells die. Release of nutrients through microbial decomposition after the death of the algae appears to be the principle means by which N is made available to the rice crop. Some algae decomposed in 2–3 days, others withstand microbial decomposition for more than 4 weeks. About 40% of algal nitrogen is being converted to ammonia within 10 days. The algal biomass when incorporated in the soil conserves organic matter, nitrogen, phosphorus, moisture and subsequently converts the sodium clay into calcium type.

Cyanobacteria are ideal and eco friendly biological systems under tropical situation for better N fixation and conservation in rice soil. Cyanobacteria are renewable biofertilizer and they represent a self-supporting biological agent because they can photosynthetically provide energy for nitrogen fixation. These Cyanobacteria may be better exploited for harnessing atmospheric nitrogen so as to increase the productivity of rice besides maintaining the soil fertility.

Zinc Solubilizing Bacteria (ZSB)

Zinc (Zn) is a key micronutrient, required for all living forms including plants, humans, and microorganisms for their development. Humans and other living organisms require zinc in their lives in little amounts for proper physiological functions. Zinc is a crucial micronutrient for plants which plays various important functions in their life cycle. Zn occurs in soil as sphalerite, olivine, hornblende, augite and biotite and availability of Zn from these sources is guided by many factors.The deficiency of zinc in the soil is one of the very common micronutrient deficiencies and results in decreased crop production. Majority of the agricultural soil is either zinc deficient or contains zinc in a fixed form which is unavailable to plants, as a result reflecting zinc deficiency in plants and soils. Therefore, to solve the above problem, there is a requirement for alternative and eco-friendly technology to enhance zinc solubilization and its availability to plants. Zinc-solubilizing bacteria (ZSB) are promising bacteria to use for sustainable agriculture. ZSB have various plant growth-promoting (PGP) properties such as Zn solubilization, P solubilization, K solubilization, nitrogen fixation, and production of phytohormones like kinetin, indole-3-acetic acid (IAA), and gibberellic acid, besides production of 1-aminocyclopropane-1-carboxylate (ACC) deaminase and siderophores, hydrogen cyanide, and ammonia. ZSB secrete different organic acids that solubilize the fixed form of zinc to available form, which enhances plant growth promotion, yield, and fertility status of the soil.

Zinc-solubilizing microorganisms can solubilize zinc from inorganic and organic pools of total soil zinc and can be utilized to increase zinc availability to plants Bacteria are known to immobilize metal by precipitation and adsorption. The ability to dissolve immobilized zinc *viz*. zinc phosphate, zinc oxide and zinc carbonate in appreciable quantity is not common feature amongst the cultivable bacteria. Few Zn solubilizing bacterial genera *viz*., *Thiobacillus thioxidans, Thiobacillus ferroxidans, Acinetobacter, Bacillus, Gluconacetobacter, Pseudomonas* and facultative thermophilic iron oxidizers have been reported as zinc solubilizers.

Silicate Solubilizing Bacteria (SSB)

Silicon is a beneficial plant nutrient with a vital role in maintaining plant growth and enhancing tolerance to both biotic and abiotic stresses. Organic weathering residues arising out of decomposition of residues, root exudates and / or metabolites from rhizosphere and non-rhizosphere microorganisms might result in the breakdown of silicate minerals releasing the ionic component.The polymeric insoluble silica present in soils is solubilised during weathering to release monosilicic acid into the soil solution which is the bioavailable form of silicon absorbed by plants. Silicon is absorbed by plant roots as silicic acid, and once absorbed it travels swiftly to active growing points, where it combines with organic compounds to improve the strength of cell walls, which in turn contribute to stronger plants.

Microorganisms are known to play major role in dissolution of minerals like silicates. Bacteria solubilizing silicates occur in the soil environment. The natural occurrence of silicate solubilizing bacteria (SSB) in rice ecosystem, in agro inputs and in materials of anthropogenic activities was observed.The organic acids and polysaccharides of microbial origin were supposed to create conditions favourable for silica solubilization. The solubilization of both insoluble silicon and phosphate by microbes is known to enhance their availability to plants. Number of bacterial strains of genus *Bacillus, Pseudomonas, Proteus, Rhizobia, Burkholderia*, and *Enterobacter* are known to release silicon from silicates and promote plant growth.

26

Mycorrhiza

Mycorrhizal symbiosis is the most prevalent and important mutualistic association in plant kingdom. 'Mycorrhiza'- the word means fungus - root, this association was reported by A.B.Frank in the year 1885. Nearly 95% of plant species form this mycorrhizal association. This serves as critical linkage between plant roots and soil and characterized by the movement of plant produced carbon to fungus and fungal acquired nutrients to plants. Through its hyphae this fungus absorbs the available P from soil and transports directly to root cortex. Mycorrhizal associations vary widely in structure and function. There are seven types of mycorrhizae (Table 1). These are endo (arbuscular), ecto, ectendo, arbutoid, monotropoid, ericoid, and orchidaceous mycorrhizae. Among them, endomycorrhizae and ectomycorrhizae are the most abundant and widespread

Table 1 : Major types of mycorrhizae and their attributes.

Mycorrhizal Type	Fungal Taxa	Plant Taxa	Intracellular Colonization	Fungal Sheath	Vesicle
Arbuscular	Glomeromycota	Bryophyta Pteridophyta Gymnosperms Angiosperms	Present	Absent	Present or Absent
Ecto	Basidiomycota Ascomycota Zygomycota	Gymnosperms Angiosperms	Absent	Present	Absent
Ectendo	Basidiomycota Ascomycota	Gymnosperms Angiosperms	Present	Present or Absent	Absent
Arbutoid	Basidiomycota	Ericales	Present	Present or Absent	Absent
Monotro-poid	Basidiomycota	Monotro-poideae	Present	Present	Absent
Ericoid	Ascomycota	Ericales Gymnosperms	Present	Absent	Absent
Orchi-daceous	Basidiomycota	Orchids	Present	Absent	Absent

Ectomycorrhizal Fungi

Amanita sp.

Ectomycorrhizas (sometimes termed ectotrophic mycorrhizas) are characteristic of many trees in the cooler parts of the world - for example pines, spruces, firs, oaks, birches in the Northern Hemisphere and eucalypts in Australia. However, some trees (e.g. willows) can have both ectomycorrhizas and arbuscular mycorrhizas, and most tropical trees have only arbuscular mycorrhizas.The fungi involved are mainly Ascomycota and Basidiomycota, including many that produce the characteristic toadstools of the forest floor. Most of these fungi can be grown in laboratory culture but, unlike the wood-rotting fungi, they are poor degraders of cellulose and other plant wall materials. So they gain most of their sugars from the living plant roots in natural conditions.

Among the different mycorrhizas, the ectomycorrhizal fungi occurring in about 10% of the world's flora include most of the gymnosperms especially the conifers and some of the angiosperm species like *Eucalyptus*, *Casuarina, Acacia*, *Betula.* The following ectomycorrhizal fungi were found in the different plantations: *Amanita muscaria, Pisolithus sp.*, *Boletus* sp., *Laccaria laccata, Lycoperdon perlatum, Rhizopogon luteolus, Russula* sp., *Suillus brevipes, S. subluteus, Scleroderma citrinum* and *Thelephora terrestris.*

Ectomycorrhizal fungi usually form a mantle that encloses the plant rootlet, from which hyphae (*i.e.*,vegetative, non-reproductive, thread like filaments) radiate outward into the soil, as well as inward between the root cells to form a hyphal network called the"Hartig net." The hyphae of most ectomycorrhizal fungal species proliferate in the duff layer of the forest floor, but some also inhabit mineral soil, and still others prefer decaying wood as a substrate. However, the plant also can benefit from the association, because the fungal hyphae that ramify into soil are very efficient in capturing mineral nutrients; these accumulate in the sheath but at least some minerals are transferred to the plant, presumably from the Hartig net.

Some ectomycorrhizal fungi require large amounts of carbohydrate, which they acquire from their plant hosts, and so are dependent on mature trees that can meet their carbohydrate demands. Although all ectomycorrhizal fungi produce spores, only some of these species able to colonize from them. Spores of mushrooms are dispersed by wind and thus can reach the centre of burns and clear cuts. Spores of truffles are dispersed primarily by small mammals. Once trees are harvested, the ectomycorrhizae remaining in the soil appear to take 1-2 years to die. Dying ectomycorrhizal roots sometimes support sufficient live hyphae to colonize new roots, depending on the fungal species involved. Thus, they are a possible source of inoculum following clear cut harvesting. Sclerotia are produced by relatively few ectomycorhizal fungal species (e.g., *Cenococcum geophilum*) sclerotia can persist from a few weeks to months, depending on the fungal species, and can be especially abundant following fire. Sclerotia contain sufficient hyphae to act as inoculum.

The Functional Role of Ectomycorrhizal Fungi

Ectomycorrhizal fungi play an important role in tree growth. It has shown that several ectomycorrhizal fungi can degrade proteins (they release protease enzymes) and thus can obtain nitrogen from the decomposing leaf litter. This could be highly significant in temperate and sub-boreal forests, where the rates of mineral nutrient recycling are low because of the low microbial activity in cool, acidic conditions. Ectomycorrhizal fungi might thus play a key role in the nitrogen nutrition of trees. It has also been shown that young seedlings growing in the shade of 'mother' trees can be attached to the 'parent' by a common network of mycorrhizal hyphae in the soil. They provide numerous benefits to their host plant with different fungal species. These include:

- enhancing the uptake of essential nutrients (mainly phosphorus and nitrogen) and water;
- protecting against pathogens and heavy metals;
- binding soil particles to create favourable soil structure;
- facilitating below-ground nutrient transfer among plants and
- altering the competitive relationships among plants of different species.

One type of ectomycorrhiza, known as a tuberculate ectomycorrhiza, also harbours nitrogen-fixing bacteria.

Endomycorrhizae

Roots of most crop plants live in association with certain beneficial fungi, the association being called mycorrhiza. Vesicular arbuscular mycorrhiza (VAM), now renamed as Arbuscular mycorrhiza (AM). It has widespread distribution throughout the plant kingdom and geographically ubiquitous (Arctic, Temperate, Tropical) in nature. About 80% of the land plants including agronomically important crops form this type of mycorrhiza.It is prevalent in most of the Angiosperms ,Gymnosperms, Pteridophytes and Bryophytes but absent in Pinaceae, Betulaceae, Ericaceae, Orchidaceae.

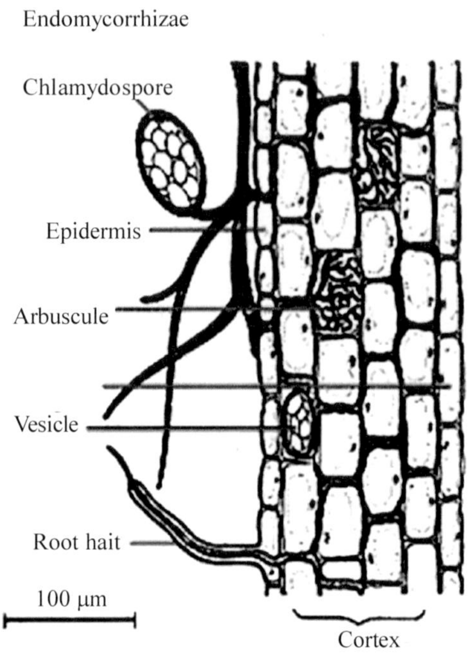

AM fungi belong to the Class Zygomycetes; Order Glomales; Family Endogonaceae. Based on spore morphology, seven genera has been characterized *viz., Glomus, Gigaspora,Acaulaspora, Entrophosphora, Scutellospora, Sclerocystis* and *Endogone.* None of these fungi has yet been successfully cultured axenically.

Components of AM Associations

Hyphae - these are non-septate when young and ramify within the cortex.

Arbuscules - intricately branched haustoria in cortex cells.

Vesicles - storage structures formed by many fungi.

Structures in Soil

Hyphae - A network of hyphae forms in the soil with thicker hyphae which function as conduits and thin branched hyphae which are thought to absorb nutrients.

Spores - Large (for a fungus) asexual spherical structures (20-1000 μm diameter) that form on hyphae in soil or roots.

Structures and Developmental Stages

AM associations form when host roots and compatible fungi are both active in close proximity and soil conditions are favourable.

Soil Hyphae

Soil hyphae, also known as extraradical or external hyphae, are filamentous fungal structures which ramify through the soil. They are responsible for nutrient acquisition, propagation of the association, spore formation, etc. AM fungi produce different types of soil hyphae including thick "runner" or "distributive" hyphae as well as thin "absorptive" hyphae. The finer hyphae can produce "branched absorptive structures" (BAS) where fine hyphae proliferate. Hyphae of AM fungi have been observed to proliferate in soil microsites where nutrients are concentrated. Hyphae of *Scutellospora* and *Gigaspora* species produce clustered swellings with spines or knobs called auxiliary cells (bodies or vesicles).

Root Contact and Penetration

Mycorrhizal associations start when soil hyphae respond to the presence of a root by growing towards it, establishing contact and growing along its surface. Next, one or more hyphae produce swellings called appressoria between epidermal cells. Root penetration occurs when hyphae from the appressoria penetrate epidermal or cortical cells to enter the root. These hyphae cross the hypodermis (through passage cells if these are present in an exodermis) and start branching in the outer cortex

Hyphal Proliferation in the Cortex

Aseptate hyphae spread along the cortex in both directions from the entry point to form a colony. The infection mycelium spread between and through cortical cells.The hyphae on reaching inner cortex can grow in to the cells and by repeated dichotomous branching form tree like structure called **arbuscules**, which facilitates inter change of metabolites between host plant and fungus.

Arbuscules

Arbuscules are intricately branched haustoria that formed within a root cortex cell. Arbuscules are formed by repeated dichotomous branching. Arbuscules start to form approximately 2 days after root penetration. They grow inside individual cells of the root cortex, but remain outside their cytoplasm, due to invagination of the plasma membrane. Arbuscules are considered the major site of exchange between the fungus and host.This assumption is based on the large surface area of the arbuscular interface. Arbuscules are short-lived and begin to collapse after a few days, but hyphae and vesicles can remain in roots for months or years.

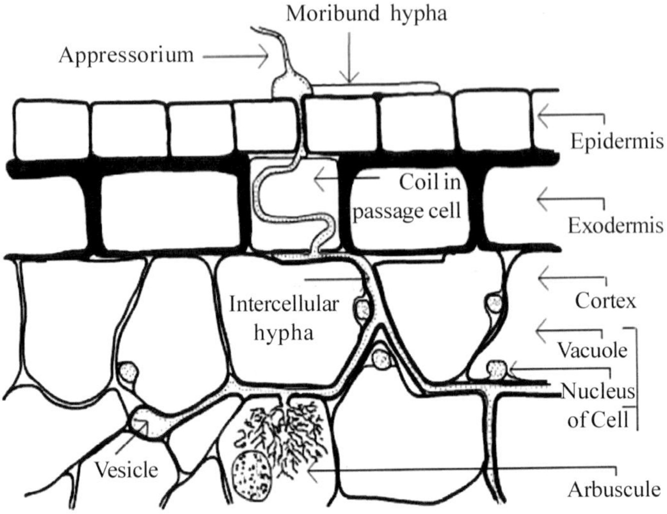

Vesicles

The extramatrical hyphae grow along the root surface, forming more penetration points and also grow out in to the surrounding soil.The fungus forms oval to spherical structures called **vesicles** with storage function. Vesicles develop to accumulate storage products in many AM associations. Vesicles are initiated soon after the first arbuscules, but continue to develop when the arbuscules senesce. Vesicles are hyphal swellings in the root cortex that contain lipids and cytoplasm. Some fungi produce vesicles which are similar in structure to the spores they produced in soil, but in other cases they are no vesicles.

Plants with mycorrhiza are called as mycorrhizal plants and plants without mycorrhiza are called as non-mycorrhizal plants.Mycorrhiza provide the host plant with mineral nutrients and water, in exchange for photosynthetic products. The AMF mycelium that emerges from the root system can acquire nutrients from soil volumes that are inaccessible to roots. Further, fungal hyphae are much thinner than roots and are therefore able to penetrate smaller pores. Carbohydrates and mineral nutrients are then exchanged inside the roots across the interface between the plant and the fungus. Thus, AMF can alleviate the limitation in plant growth caused by an inadequate nutrient supply. In addition to an improved nutritional supply, AM interactions provide other benefits to plants, such as improved drought and salinity tolerance and disease resistance. AM fungi are known to alleviate heavy metal toxicity in the host plants and to tolerate high metal concentrations in the soil. AM fungi can also have a direct effect on the ecosystem, as they improve the soil structure and aggregation and drive the structure of plant communities and productivity.

The enhanced growth of plants has been attributed to nutritional and non-nutritional effects of AM fungi. It has been reported to benefit plants by increasing the uptake of nutrients such as P, Zn, Cu, and nitrogen. The non nutritional effects of AM would be due to increased tolerance to saline conditions, improved water relations increased survival ratio of transplanted seedlings, control of root diseases and increased soil aggregation by the external hyphal net work. The Mycorrhizal fungi are also known to produce wide array of plant growth promoting substances (IAA, IBA, GA).

Like roots, mycorrhizal fungal hyphae also release exudates containing carbon into the surrounding soil, the 'hyphosphere' that contributes to the formation of microbial communities and aggregation of soil particles. The soil region influenced by the mycorrhizal roots is the 'mycorrhizosphere'. A wide range of microorganisms like bacteria, fungi, protozoa, nematodes, arthropods, etc., inhabit the mycorrhizosphere. These microorganisms interact with each other and with the plant system either directly or indirectly. The activities in the mycorrhizosphere include stimulation in the activities and populations of microorganisms, changes in pH, nutrient release from organic matter and nutrient cycling, suppression of plant pathogens, mycorrhizal formation, and changes in soil structure.

The most important feature common to mycorrhizal fungi are their ability to recognize the roots of appropriate host plants to infect root tissues without eliciting rapid or intense host defense reactions. The mycelium produced by mycorrhizal fungi can

i. Survive as resting propagule in the absence of host
ii. Grow independently in soil
iii. Grow as symbiotic mycelium in soil
iv. Grow in close association with roots
v. Form reproductive organs.

These fungi in addition to supplying phosphorus to the plant offer protection against root diseases. They also help in water uptake and overcoming transplant shock thus help in better establishment of plants.

AM fungi on Phosphorus Contribution

Phosphate transport by AM is a key factor in plant nutrition. AM fungi can absorb several times soil P and have greater P inflow rates than roots. The fungal hyphae actually transport P over large distances into the root. The phosphate absorbed by VAM fungi from soil solution is accumulated in the vacuoles of the fungi as poly phosphate granules and this represents 16-40% of the total P in AM fungi. The poly beta granules are propelled through the hyphae by cytoplasmic streaming to the arbuscules. The inflow rate through external hyphae is approximately thousand times faster than the diffusion rate through the soil pores. It is also important that when phosphate supply is in excess, polyphosphate accumulates the nutrient as a storage pool that can be used at times of increasing demand to support stages of plant growth. Increased uptake of P due to inoculation with AM fungi in various agricultural crops has been reported.Mycorrhizal fungi also secrete enzymes, such as phosphomonoesterases and phosphatases that are able to hydrolyze organic P compounds in soil, such as phenolphthalein diphosphate, providing plant hosts with phosphorous typically unavailable to non-mycorrhizal roots.

AM fungi on Nitrogen Contribution

Nitrogen bound in organic matter is present as peptides, proteins, and free amino acids, which can be directly absorbed by fungi. Both arbuscular mycor-rhizal and ectomycorrhizal fungi release peptidases and proteases into the soil that cleave organically bound nitrogen and subsequently absorb the nitrogenous monomers.Inorganic nitrogen in the soil is preferentially taken up by fungi as nitrates (NO_3^-) or ammonium (NH_4^+), and it has been suggested that, once this nitrogen has been absorbed by the extraradical mycelia, it is converted into the amino acid,arginine for transport into the intraradical mycelia. Once internal migration is complete, the arginine would be broken down through the urease cycle into ammonium for transport into the plant.Arginine is the final product utilized by plants.

Role of AM fungi in Water Uptake

In addition to nutrient uptake, mycorrhizae also enhance the transport of water from soil to plant. Since mycorrhizal plants have higher transpiration rates, the water uptake per unit root length and thus the massflow of soil solution to the root surface is about twice as high as in non mycorrhizal plants. The external mycelium of AM fungi play an important role in water uptake of plants, particularly in semiarid habitats probably for not assuring high transpiration rates but for preserving physiological activities. Water uptake contribution through the external mycelium become more beneficial for the host plant in water stress condition. It has been reported that mycorrhizal associations help plants to increase nutrient uptake during water-stressed conditions by increasing hydraulic conductivity in roots.

Rolc of AM fungi on Biological Control

The plants inoculated with AM fungi exhibit increased resistance to fungal root diseases like wilts and root rots. The severity of nematode disease is generally reduced in mycorrhizal plants. Growth responses due to AM fungi may alter disease resistance in shoot and leaf. Several results reported the increased tolerance of AM plants to root pathogens.Several mechanisms were proposed for the suppression of the pathogens as physical barrier, improved P nutrition of the host plant, stimulation natural microbionts, competition, induction of isoflavanoids, tissue lignification, induction of chitinase and peroxidases.

With AM formation, production of plant defense chemicals like phenolic substances, phytoalexins, and chitinases are increased. Symbiotic processes are not affected by these chemicals but systemic plant defense mechanisms are turned on which in turn defend host plants from other harmful pathogens. Competitive inhibition of pathogens by endomycorrhizal fungi is demonstrated to protect host plants from diseases like root rot, collar disease, etc.

Other Benefits of AM Fungi

- AM fungi enhances the uptake of Zn, Cu, K, Ca, Mg.
- It increases plant growth due to growth hormone production.
- It increases nodulation and N_2 fixation.
- It offers resistance to diseases and nematodes.
- AM increases drought tolerance.
- It saves 25 % P fertilizer & increases yield by 10 -15 %.
- It increases P use efficiency.
- It increases microbial activity in mycorrhizosphere.

Ectendomycorrhiza

The ectendomycorrhiza is confined to *Pinus* and *Larix* spp. and is common in conifer nurseries and in disturbed habitats. This association is characterized by the unique combination of a fungal mantle, Hartig net, and intracellular hyphae. *Sphaerosporella brunnea, Wilcoxina mikolae* fungi have been reported to form ectendomycorrhizas.

Orchid Mycorrhiza

These are endomycorrhizas formed between plants of the Orchidaceae and basidiomycetous or rarely ascomycetous fungi. The fungi form intracellular coils or less regular hyphal aggregates within host tissue. These structures are known as 'pelotons'.For their life cycle, orchids are obligately dependent on their mycorrhizal partner.Some orchids cannot photosynthesize prior to the seedling stage. Other orchids are entirely non-photosynthetic. All orchids, however, depend on the sugars provided by their fungal partner for at least some part of their lives. *Armillaria* and *Rhizoctonia* are colonizing the orchid plants.Orchid seeds require fungal invasion in order to germinate because, independently, the seedlings cannot acquire enough nutrients to grow. Under natural conditions, orchid seeds will not successfully germinate without infection by the mycorrhizal fungus. Once the seed coat ruptures and roots begin to emerge, the hyphae of orchidaceous mycorrhiza penetrate the root's cells and create hyphal coils, or pelotons, which are sites of nutrient exchange.

Eg. Orchid: *Goodyera repens* Mycorrhizal fungi: *Rhizoctonia goodyerae*

Ericaceous Mycorrhiza

Ericaceous mycorrhizae are generally found on plants of the order Ericales and in inhospitable, acidic environments. It is associated with the members of Ericaceae family (*Rhododendron, Azalea, Vaccinium*) plants. The fungi belong to Ascomycetes and they are cultivable. *Hymnoschypus ericae*, colonizing healthy plants like straw berry and blue berry.While they do penetrate and invaginate the root cells, ericoid mycorrhiza do not create arbuscules. They help to regulate the plant's acquisition of minerals including iron, manganese, and aluminum. Additionally, mycorrhizal fungi form hyphal coils outside of the root cells, significantly increasing root volume. Another well known example is *Pezizella ericae.*

Arbutoid Mycorrhiza

Arbutoid mycorrhiza are a type of endomycorrhizal fungi that look similar to ectomycorrhizal fungi. Arbutoid Mycorrhizae are found in the plant genera *Arctostaphylos* and *Arbutus*. The fungi that form arbutoid mycorrhizal relationships are basidiomycetes. In arbutoid mycorrhizal associations a fungal sheath, or mantle covers the roots of the host plant, similar to the structure of ectomycorrhizal associations. The fungal hyphae also form a structure known as a hartig net into the outer cortical cells. Nutrients acquired from the soil pass through the mantle into the roots of the host plant. A major difference between the arbutoid and ectomycorrhizal association is that the hyphae of the arbutoid penetrate the outer cortical cells, and fill them with coils, which the hyphae of ectomycorrhizas do not. The intracellular coils, along with the mantle sheath and Hartig net are the diagnostic features of arbutoid mycorrhizas. Eg. *Arbustus* and *Pyrola*

Monotropoid mycorrhiza

The mycorrhizal association formed by plants of the Montropaceae are called as Monotropoid. The Monotropaceae are also unusual, in that all ten genera in the family are entirely achlorophyllous. This means they contain no chlorophyll, and hence are unable to photosynthesise and produce carbohydrates. Instead,

27

Biocontrol Agents

For many soil borne diseases, effective and economic control methods are not available. Several soil borne pathogens have developed resistance to many fungicides. The beneficial microbes are also affected by fungicides. Many fungicides pose health hazards by polluting food and water and cause direct toxicity of heavy metals and carcinogenic risk. In the context of growing worldwide awareness of environmental pollution and development of fungicide resistant strains of pathogens, biological method of suppression of disease causing organisms have been recognized as the safest approach for plant disease management.

Biological control of plant diseases is the suppression of populations of plant pathogens by living organisms. Amongst beneficial microorganisms, isolates can be selected which are highly effective against pathogens and can be multiplied on artificial media. Application of such selected and mass produced antagonists in high densities once or several times during a growing season is called "augmentative biological control".

An ideal biocontrol agent should have the following features/characteristics.

1. It should not be pathogenic to plants, human beings, animals and microorganisms.
2. It should have broad spectrum of activity in controlling many types of diseases and must be genetically stable.
3. It should have fast growth and sporulation.
4. It must be cultured under artificial media.
5. The inoculum must be capable of abundant production using conventional methods such as liquid fermentation and withstand long term storage until application.
6. It should be amenable for inexpensive mass multiplication and economically viable.
7. It should have long shelf life.

8. It must be efficacious under different environmental conditions.
9. It should be compatible with biofertilizers.
10. It should have least susceptibility to the action by the seed treating chemicals.
11. It should not be toxic to beneficial organisms in or on the target area.
12. It should be easily formulated and methods of application must be convenient with common cultural practices.
13. It should easily establish in the soil with high persistence and survival capacity.
14. It should be biologically competitive with other microorganisms.

Biocontrol Agents

Biocontrol Agents		Diseases controlled
Bacterial antagonists		
Agrobacterium radiobacter	-	*Agrobacterium tumefaciens*
Bacillus subtilis	-	*Fusarium, Rhizoctonia, Gaeumannomyces*
Pseudomonas fluorescens	-	Blast & sheath blight of paddy, wilt of Banana, Redgram and Chickpea.
Pseudomonas cepacia		
Pseudomonas solanacearum		
Fungal antagonists		
Trichoderma viride	-	*Fusarium, Phytophthora, Pythium, Rhizoctonia, Macrophomina*
Trichoderma harzianum	-	*Pythium, Gaeumannomyces, Armillaria, Fusarium, Rhizoctonia*
Trichoderma hamatum	-	Dampingoff, root rot.
Trichoderma virens	-	Dampingoff, root rot.
Actionomycetes		
Streptomyces griseoviridis	-	*Alternaria brassicae, Fusarium oxysporum*
Virus		
Tomato mosaic virus (mild strain)	-	Tomato mosaic virus (severe strain)
AM (Arbuscular Mycorrhiza) Fungi		
Glomus fasciculatum		
Glomus aggregatum		Root rot, wilt, dampingoff, stem rot and sheath blight of paddy
Glomus mosseae		
Glomus etunicatus		

Mechanisms Involved In Biological Control

The biocontrol agents control the plant diseases by the following general mechanisms,

a. Competition

b. Antibiosis

c. Parasitism

d. Lysis

a. Competition

Competition between antagonist and pathogens as a mechanism of biocontrol may occur at different levels. For example, competition for space or specific infection sites on roots or seeds has been proposed as a mechanism of biocontrol of pathogenic *F. oxysporum* by nonpathogenic strains of *F. oxysporum* and of pathogenic strains of *R. solani* by nonpathogenic *Rhizoctonia* spp. Competition between microorganisms for carbon, nitrogen, and other nutrients in the rhizosphere is another well established mechanism of biocontrol. Competition for iron, mediated by production of iron-chelating siderophores, has been demonstrated as a mechanism of biocontrol by several species of bacteria in soils where iron is limiting. This is a widely recognized mechanism of biocontrol by fluorescent *Pseudomonas* spp., which produce a range of siderophores including pseudobactins and pyoverdines. The siderophores are thought to sequester the limited supply of iron that is available in the rhizosphere to a form that is unavailable to pathogenic fungi and other deleterious microorganisms, thereby restricting their growth.

b. Antibiosis

It is defined as inhibition of an organism or its death by the action of toxic metabolites (antibiotics) produced by another organism. Antibiotics are generally considered to be organic compounds of low molecular weight produced by microbes. One or more antibiotics play a role in disease suppression in biocontrol system. Antagonistic bacteria produced many antifungal metabolites like, 2, 4 diacetylphloroglucinol (DAPG), HCN, butyrolactones, kanosamine, oligomycin A, oomycin A, phenazine -1- carboxylic acid, pyoluteorin, pyrrolnitrin, viscosinamide, xanthobaccin and zwittermycin A. Likewise *Trichoderma* spp.produce larger quantities of fungistatic metabolites like trichodermin, dermadin, trichoviridin, sesquiterpene, hepatalic acid, trichobrachin, gliotoxin and gliovirin which are active against many pathogens/diseases.

Trichoderma sp. produce number of antibiotic compounds like Trichodermin, Trichoviridin and viridol. *Pseudomonas fluorescens* produce antibiotics called 'pyoluteorin' which kills *Pythium ultimum* and 'pyrol nitrin' kills *Rhizoctonia solani*.

c. Parasitism

Parasitism of plant pathogens as a mechanism of biocontrol is usually associated with fungal biocontrol agents. Most evidence for this comes from field observations of infected fungal propagules such as spores or sclerotia. For example, oospores of *Phytophthora* and *Pythium* spp. are frequently found to be infected by *Olpidiopsis gracilis*, whilst sclerotia of *R. solani* are infected by the obligate sclerotial mycoparasite *Verticillium*. The interaction between the mycoparasite and its host involves a sequence of processes encompassing contact, recognition, localized lysis, penetration, intracellular growth, and exit. Mycoparasites secrete different cell wall-degrading enzymes (e.g., β-1,3-glucanases, chitinases, proteinases, and lipases) during the penetration process. Other mechanisms of parasitism are associated with fungi such as *Verticillium* and *Paecilomyces*, which can infect the egg masses and cysts of the cereal cyst and root knot nematodes.

d. Lysis

Lytic enzymes play a vital role in mycoparasitism and degrade the pathogen cell wall. The enzymes like chitinases and β-1, 3 glucanases lyse the host cell wall and lead to the leakage of protoplasmic contents which are in turn used as a food material for the multiplication of the antagonist. Different fungi and bacteria secrete hydrolytic enzymes and degrade the cell wall of the pathogens & causes lysis.

Eg: *Trichoderma* sp. produces β-1,3 glucanase and chitinase, which degrade chitins and glucans in cell wall and causes its lysis in *R. solani, Pythium* sp. etc.,

Other mechanisms in biological control are more specifically host-related/ restricted. They include:

i. Cross protection

ii. Hypovirulence

iii. Induced systemic resistance.

i. Cross protection

It is the phenomenon in which plants systemically infected with one strain of a virus are protected from infection by a second related severe strain of the same virus. This phenomenon has been reported in different crops.

The plant diseases controlled by using cross protection are :

- Papaya ring spot virus (PRV)
- Tomato mosaic virus (TMV)
- Citrus tristeza virus (CTV)
- Apple mosaic virus (ApMV)
- TMV and Cucumber mosaic virus (CMV)
- Soybean mosaic virus (SMV)

The phenomenon of cross protection has been established in some of the bacterial diseases in plants. A plant infected with a mild strain of the bacterium escapes disease from infection by severe strain(s). Avirulent strain of *A.tumefaciens* protect the apple crop from severe infection by virulent strains of *A.tumefaciens*. The same mechanism has been reported in apple fire blight bacterium (*Erwinia amylovora).* Heat killed cells of *Pseudomonas tabaci* protected tobacco crop from angular leaf spot or wild fire by an avirulent strain.

ii. Hypovirulence

Inoculation of diseased plants with hypovirulent strains carrying double-stranded RNA / DNA plasmids can result in reduced virulence of pathogen strain.

Eg: Hypovirulent strain of *Rhizoctonia solani* - to control pathogenic strain of *R.solani*

iii. Induced Systemic Resistance (ISR)

Induced systemic resistance (ISR), also referred to as systemic acquired resistance, refers to the situation where the plant acquires resistance to infection from a pathogen.The biocontrol agents bring about induced systemic resistance (ISR) through fortifying the physical and mechanical strength of cell wall as well as changing physiological and biochemical reaction of host leading to synthesis of defense chemicals against pathogens. Development of ISR is generally accompanied by the expression of a set of genes within the plant, including those that encode for pathogenesis-related proteins (PR proteins) such as chitinases and β-1,3-glucanases. Defense reaction occurs due to accumulation of PR proteins (chitinase, β-1, 3 glucanases), chalcone synthase, phenylalanine ammonia lyase (PAL), peroxidase, phenolics, callose, lignin and phytoalexins. Chitinolytic enzymes together with β-1,3 glucanases or cellulases are most frequently considered to play a vital role in biocontrol.

Management of plant diseases by Biocontrol agents

Crop	Disease	Biocontrol agents used	Method of application
Rice	Sheath blight	*P. fluorescens.* *Bacillus subtilis*	Seed treatment, soil application and foliar spray
	Sheath rot	*Trichoderma* spp. *P. fluorescens*	Seed treatment
	Brown leaf spot	*Trichoderma* spp. *P. fluorescens*	Seed treatment
	Blast	*P. fluorescens* *Bacillus subtilis*	Seed treatment and foliar spray
	Bacterial blight	*P. fluorescens.,* *Bacillus subtilis*	Seed treatment and foliar spray
Wheat	Loose smut	*P. fluorescens*	Seed treatment
Cumbu	Ergots	*Trichoderma* spp.	Seed treatment
Pulses	Dry root rot	*Trichoderma* spp.	Seed treatment and Soil application
Chickpea	Wilt	*Trichoderma* spp.	Seed treatment
Pigeonpea	Wilt	*Trichoderma* spp. *Bacillus subtilis*	Seed treatment
Peanut	Dry root rot	*Trichoderma* spp.	Seed treatment and soil application
	Stem rot	*Trichoderma* spp.	Seed treatment and soil application
	Collar rot	*Bacillus subtilis*	Seed treatment
Sesame	Root rot	*Trichoderma* spp. *Bacillus subtilis*	Seed treatment and soil application
Sunflower	Charcoal rot	*Trichoderma* spp.	Seed treatment and soil application
Cotton	Root rot	*Trichoderma* spp. *Bacillus subtilis*	Seed treatment and soil application
	Wilt	*Trichoderma* spp. *Bacillus subtilis*	Seed treatment and soil application
	Bacterial blight	*P. fluorescens,* *Bacillus subtilis*	Seed treatment and soil application
Sugarcane	Damping off	*Trichoderma* spp. *P. fluorescens*	Seed treatment and foliar spray
Tobacco	Damping off	*Trichoderma* spp. *P. fluorescens*	Seed treatment
Tomato	Damping off	*Trichoderma* spp. *P. fluorescens*	Seed treatment
	Wilt	*Trichoderma* spp.	Seed treatment, Seedling dip and soil application

Brinjal	Damping off	*Trichoderma* spp. *P. fluorescens*	Seed treatment
	Wilt	*Trichoderma* spp.	Seed treatment and soil application
Watermelon	Wilt	*Trichoderma* spp.	Seed treatment and soil application
Potato	Late blight	*P. fluorescens*	Foliar spray
Banana	Anthracnose	*Trichoderma* spp. *P. fluorescens*	Fruit dip
	Fungal wilt	*P. fluorescens*	Sucker Treatment, soil application and capsule application
	Bacterial wilt	*P. fluorescens*	Sucker Treatment, soil application and capsule application
Mango	Anthracnose	*Trichoderma* spp. *P. fluorescens*	Fruit dip
	Bacterial canker	*P. fluorescens*	Foliar spray and fruit dip
Guava	Anthracnose	*P. fluorescens.,* *Bacillus subtilis*	Fruit dip
	Fruit rot	*Bacillus subtilis.* *P. fluorescens*	Fruit dip
Citrus	Green mold	*Bacillus subtilis.* *P. fluorescens*	Fruit dip
	Canker	*Bacillus subtilis*	Foliar spray
Coconut	Wilt	*P. fluorescens*	Soil application
Chilli	Fruit rot	*Trichoderma* spp. *P. fluorescens*	Seed treatment and seedling dip
	Damping off	*Trichoderma* spp. *P. fluorescens*	Seed treatment and seedling dip
Coriander	Wilt	*Trichoderma* spp.	Soil application
Turmeric	Rhizome rot	*Trichoderma* spp. *P. fluorescens*	Rhizome treatment
Ginger	Rhizome rot	*Trichoderma* spp. *P. fluorescens*	Rhizome treatment
Black pepper	Collar rot	*P. fluorescens.,* *Bacillus subtilis*	Foliar spray
Betelvine	Foot rot	*Trichoderma* spp. *P. fluorescens*	Soil application
Jasmine	Root rot	*Trichoderma* spp.	Cutting treatment
Crossandra	Wilt	*Trichoderma* spp.	Soil drench

Advantages of Biocontrol Agents

- Biocontrol agents are less costly and cheaper than any other methods.
- Biocontrol agents give protection to the crop throughout the crop period.
- Biocontrol agents not only controls the diseases but also enhance the growth by way of encouraging the beneficial soil microflora.

- Application of biocontrol agents is safer to the environment and the person applying them.
- They are highly effective against specific diseases.
- It can eliminate pathogens from the site of infection.
- They are very easy to handle and apply to the targets.
- Biocontrol agents can be combined with bio fertilizers.
- They do not cause toxicity to the plants.
- They multiply easily in the soil and leave no residual problems.

Disadvantages of Biocontrol Agents

- Biocontrol agents can be used only against specific diseases.
- They are less effective than the fungicides.
- Few biocontrol agents are available for use and are available only in few places.
- They are unavailable in larger quantities at present.
- They have slow effect in the control of plant diseases.
- This method is only a preventive measure and not curative.
- Shelf life of biocontrol agent is short. *Trichoderma viride* - 4 months *Pseudomonas fluorescens*- 3 months
- Efficiency of biocontrol agents is decided by the climatic conditions.
- The required amount of population of biocontrol agents should be checked at periodical intervals and should be maintained at required level.
- Certain circumstances a biocontrol agents may become a pathogen.

Methods of Application of Biocontrol Agents

Fungi like *Trichoderma viride*, and bacteria like *Bacillus subtilis* and *Pseudomonas fluorescens* are commonly used as biocontrol agents in the management of plant diseases.

Methods of application of biocontrol agents

i. Seed treatment

ii. Seedling root tip

iii. Sucker treatment

iv. Capsule application

v. Soil application

vi. Foliar spraying

i. Seed treatment

- *Trichoderma viride* – dry seed treatment – 4g/kg of seeds, reduce the root rot diseases of black gram, green gram, sesame, cotton, sunflower and groundnut.
- *Pseudomonas fluorescence*- 10g/kg of seeds reduce the infection of rice blast & sheath blight, wilt of pigeon pea and chick pea.

ii. Seedling root dip

- Irrigation water should be stagnated in an area of 25sq.m in rice field.
- 2.5kg of *Pseudomonas fluorescence* should be mixed with the stagnated water.
- Seedlings are to be immersed in the stagnated water for a minimum period of 30 min. and transplanted. Eg. rice blast.

iii. Sucker treatment

- 10g of *Pseudomonas fluorescens* is sprinkled on clay dipped suckers before planting to control panama wilt of banana.

iv. Capsule application

- Capsule application with *P. fluorescens* is used to manage panama wilt of banana.
- Each capsule is filled with 50mg of *P. fluorescens* and applied @ one capsule to each banana sucker in the field starting from third month of planting and repeated twice at monthly intervals for effective control.

v. Soil application

Trichoderma viride

- Talc based formulation of *Trichoderma viride* (2.5kg) should be mixed with well decomposed FYM or sand (50kg) and then applied to soil (@ 2.5kg/50kg FYM/ha).

- Recommended for root rot of pulses and oil seeds & wilt of pigeon pea and chick pea.

Pseudomonas fluorescens

- *P. fluorescens* (2.5kg) formulation is mixed with well decomposed FYM or sand (50kg) and applied (@ 2.5kg/50kg FYM/ha)
- Recommended for rice blast & panama wilt of banana.

vi. Foliar spraying

P. fluorescens is recommended for the control of blast of rice at 0.2% concentration (2kg/ha). The product is to be sprayed 3 times from 45 days after transplanting at 10 days intervals.

28

Biopesticides

Chemical pesticides though extremely effective in controlling pest population suffer from disadvantages. They have a broad toxic range and potentially toxic to non target organisms. Other constraints that have emerged with the use of chemical pesticides such as development of resistance and resurgence in insects, residual toxicity, secondary outbreak of minor insect pests, etc., In this juncture, the biopesticides have proven to be effective tool in managing the devastating insect pests of various crops.

The biopesticides are types of pesticides derived from natural materials such as microorganisms, plants, animals and certain minerals. They are living organisms or their products or byproducts, which can be used for the management of pests that are injurious to crop plants. Biopesticides have a vital role in crop protection, although most commonly in combination with other tools including chemical pesticides as part of integrated pest management (IPM). Microorganisms help to regulate the abundance of many pests or potential pests without any cost to society. Microbes can be increased artificially and thereby achieve faster, higher and more consistent control of the target pests that will occur naturally. Insect pathogens that are transmissible in nature from one host to another offer an important bonus effect that chemical insecticides do not have. A pathogen will kill pests exposed and many continue killing them by natural transmission from one individual host to another through subsequent host generation.

Biopesticides generally have several advantages compared to conventional pesticides. While chemical pesticides are responsible for extensive pollution of the environment, a serious health hazard due to the presence of their residues in food and development of resistance in targeted insect pest populations, biopesticides, in contrast, are inherently less toxic to humans and the environment, do not leave harmful residues, and are usually more specific to target pests. The agents employed as biopesticides, include parasites, predetors and disease causing fungi, bacteria and viruses, which are the natural enemies of pests. Utilization of naturally occurring parasites, predators and pathogens for pest control is a classical biological control. These bio agents can be conserved, preserved and multiplied under laboratory condition for field release. Once these bio-agents are introduced in the field to build their population considerably, they are capable of bringing down the targeted pest population below economic threshold level (ETL).

Different Biopesticide Groups of Microorganisms

1. Bacterial pathogens
2. Fungal pathogens
3. Viral pathogens
4. Protozoan pathogens

Bacterial pathogens

Bacteria which affect insects are broadly classified as spore formers and non spore formers.

Spore formers

Bacillus thuringiensis, B.popilliae, B. lentimorbus, B. sphaericus,.

Non-spore formers

Serratia, Pseudomonas, Aerobacter and Streptococcus

a. *Bacillus thuringiensis (Bt)*

This is the widely exploited species of *Bacillus*. It is environmentally safe, active on insects resistant to chemical insecticide and selective against pests. *B. thuringiensis* var. *kurstaki* was isolated from *Heliothis armigera* in 1955. Bt is a gram positive soil bacterium which, during sporulation produces proteinaceous crystalline inclusions with insecticidal activity. The insecticidal crystalline proteins (Cry proteins) of different Bt strain exhibit striking difference in their insecticidal spectra and are not toxic to non target organisms.

Four major classes of Cry proteins

Cry I - Lepidoptera specific

Cry II - Lepidoptera and Diptera specific

Cry III - Coleoptera specific

Cry IV - Diptera specific

After ingestion, Bt crystalline inclusions dissolve in the midgut of insect larvae and release one or more crystalline proteins. These proteins are 'prototoxins' proteolytically converted in to smaller toxic polypeptides called endotoxins. The active endotoxins bind to the midgut epithelial cells, leading to the disruption of membrane. As a consequence, the epithelial cells lyse and the insect stop feeding and eventually die.

Bt based pesticides are being used for the last 4 to 5 decades for the control of insect pests. Cry genes are first introduced into tobacco against tobacco horn worm and subsequently in to potato and cotton. Monsanto Company developed Bt transgenic Cotton in the year 1996. Delivering Bt genes in to plants would help to control the lepidopteran larvae in cotton, corn, rice and vegetables.

There are several varieties of *Bt*

Bt. var *thuringiensis*

Bt .var *lentimorpus*

Bt .var *sotto*

Bt .var *Canadensis*

***Toxins* of Bt.**

They are of 3 different types.

i. α – exotoxin

They are water soluble, heat labile toxin, identified as lecithin.

ii. β - exotoxin

They are water soluble, heat stable toxin, highly toxic to several flies. Chemically this is adenosine nucleotide analog of ATP. Thuringiensis is a specific inhibitor of DNA dependent RNA polymerase and ATP.

iii. δ - endotoxin

This is most important one, called as crystalline toxin.

Bt and their target pests

Bt strains	Target pest
B. thuringiensis bv. *kurstaki*	Caterpillars
B. thuringiensis bv. i*sraelensis*	Mosquitoes Black flies
B. thuringiensis bv. *tenebrionis*	Colorado Potato beetle
B. thuringiensis bv. a*izawi*	Wax – moth Diamond back moth

Symptoms of bacterial infections

1. The body of the insect or larvae becomes soft or dried to a blackscale.
2. Body usually darkened in colour with dark fluids tissues disintegrated with putrefied odour.
3. Body becomes milky white in colour with body fluids white. (eg. Infection in *B. popilliae*).
4. Body becomes soft and turns to red colour (*Serratia marcesens* infection).
5. In Lepidopteran insects, insects may be found hanging by prolegs in plants or twigs.

Fungal pathogens

Entomopathogenic fungi have their ability to decimate insect host populations, particularly during epizootics. Some fungi have very broad host ranges. Some of them are :

Beauveria bassiana

Metarrhizium anisopliae

Verticillium lecanii

Paecilomyces spp.

Nomurea rileyi

a. *Metarrhizium*

It is used in the control of oryctus rhinocerous of coconut. The adult beetle causes damage to spathe of adult trees and young plants. They are multiplied near farm yard manure area. First it was named as *Entomopthora anisopliae* and latter renamed as *Metarrhizium anisopliae*. The other species is *M. flavoviride*. *M. anisopliae* was recorded in 200 insect host belonging to different groups.

- Orthoptera
- Lepidoptera
- Diptera
- Hymenoptera, etc.

This is commonly called Green muscardine fungus. When young, the mycelium is white in colour and after 4-5 days the conidia is produced, then colour changes to dark green.*M. anisopliae* generally enters insects through spiracles and pores in the sense organs. Once inside the insect, the fungus produces a lateral extension of hyphae, which eventually proliferate and consume the internal contents of the insect.Hyphal growth continues until the insect is filled with mycelia. When the internal contents have been consumed, the fungus breaks through the cuticle and sporulates, which makes the insect appear 'fuzzy'. The fungi can also produce toxins called *Destructins* A, B, C, D, E and desmethyl destruxin B, which have insecticidal properties on moth and fly larvae. Mass multiplication is carried out in grubs or in laboratory media.

b. *Beauveria*

First identified in 1912 more than 10 species are identified. Very few are reported to be effective in the pest control. *B. bassiana* is a commercially exploited. Toxins are called Beauvericins. Toxic to mosquito larvae and also to potato beetle.

Symptoms

Sluggish movement of larvae, loss of appetite, partial paralysis, reduction in capacity to lay eggs in females, discolouration – are observed initially. Most important stage is the *Sclerotium* formation stage.

Viral Pathogens

Nuclear Polyhedrosis Virus (NPV)

It is an obligate pathogen. The virus consists of proteinaceous polyhedral occlusions body inside which the virions or virus rods are embedded. The viral particles are ingested into insects body when feeding on treated plant parts. Due to alkaline gut juice, the virus are liberated which attack nuclei and cell tissue, fat bodies, tracheal matrix, haemocytes, ganglia and brain. The infected larva become dull in colour, less active and become flaccid, skin becomes very fragile and eventually ruptures. Diseased larvae hang upside down from the plants. This is called 'tree top disease' or "Wipfelkrankheit".

NPV is specific and available for *Helicoverpaarmigera* (Ha NPV), *Spodoptera litura* (Si NPV), *Amsacta albistriga* (Aa NPV)

Granulosis Virus (GV)

Virions are occluded singly in small inclusions bodies called capsules. It attacks cytoplasm or nucleus of epidermis, trachea and fat bodies. Infested larvae lose appetite become sluggish and slow milky white colour on ventral surface. It is recommended for Sugarcane early shoot borers.

Mechanism of action

They include the following

1) Antibiosis
2) Competition
3) Hyperparasitism
4) Synergism

1. Antibiosis

This occurs as a result of an interaction with other microbes mediated by specific metabolite of microbial origin, by volatile compounds, lytic enzymes or other toxic substances. The microorganisms produce antibiotics, bacteriocin, volatile compound and metabolite production.

2. Competition

Another mechanism of control by biopesticides is their ability to compete aggressively, that they grow rapidly and colonize substrate to exclude pathogens. For example *Trichoderma* sp. are aggressive competitors of *Fusarium* sp.

3. Hyperparasitism

Hyperparasitism is the lysis of the death by other microorganisms or direct parasitism. For example *T. lignorum* is found to be parasitizing the hyphae of *R. solani* and therefore soil inoculation with *Trichoderma* spores help to control damping off disease in citrus seedlings.

4. Synergism

The ability of some bioagent to combine actions of hydrolytic enzymes and antibiotic secondary metabolites. For example the effectiveness of *Trichoderma* sp. as a biocontrol agent is as a result of its synergistic effects of antimicrobial compounds *viz*., pyrones,coumarins, etc.

Advantages of Biopesticides

- Biopesticides are usually inherently less harmful and cause less environmental load or pollutions.
- Development of resistance is not a problem.
- Designed to only one specific pest or, in some cases, a few target pests as opposed to chemical that have a broad spectrum activity.
- Cost of developing biopesticides is significantly lower than those of synthetic chemical pesticides.
- Their nature of control is preventive not curative.

29

Mass Production of Bioinoculants

Mass Production of Bacterial Inoculants

Strain Selection and Improvement

In biofertilizer technology, strain selection occupies the foremost criteria, so as to explore its biological potentialities for the beneficiary crop. The selected strain used in biofertilizer production must be efficient to grow fast besides retain desirable potentialities of biological activities, which should be present in desired high numbers and must be stable. The strain must be improved periodically so as to sustain the productivity.

The mass production of carrier based bacterial biofertilizers involves three stages.

1. Culturing of microorganisms
2. Processing of carrier material
3. Mixing the carrier and the broth culture

1. Culturing of microorganisms

Although many bacteria can be used beneficially as a biofertilizer, the technique of mass production is standardized for *Rhizobium*, *Azospirillum*, *Azotobacter* and phosphobacteria (Phosphate solubilizing microorganisms).

The media used for mass culturing are as follows:

Rhizobium	:	Yeast extract mannitol broth.
Azospirillum	:	Dobereiner's Malic acid broth
Phosphobacteria	:	Nutrient glucose broth

Composition of media

Yeast Extract Mannitol broth

Mannitol	10.0 g
K_2HPO_4	0.5 g
Yeast Extract	0.5 g
$MgSO_4$	0.2 g

NaCl	0.1 g
$CaCO_3$	3.0 g
pH	7.0
Distilled water	1000.0 ml

Dobereiner's Nitrogen free malic acid broth

Malic acid	5.0 g
KOH	4.0 g
K_2HPO_4	0.6 g
KH_2PO_4	0.4 g
$MnSO_4$	0.01 g
$MgSO_4$	0.05 g
NaCl	0.02 g
Na_2MoO_4	0.002 g
Distilled water	1000.0 ml
Bromothymol blue (0.5% alcoholic solution)	2.0 ml
pH	6.6 – 7.0

Nutrient glucose broth

Beef Extract	3.0 g
Peptone	5.0 g
Glucose	5.0 g
NaCl	5.0 g
Distilled water	1000.0 ml
pH	6.5 – 7.0

The broth is prepared in flasks and inoculum from mother culture is transferred to flasks. The culture is grown under shaking conditions at 30±2°C as submerged culture. The culture is incubated until maximum cell population of 10^{10} to 10^{11} cfu/ml^{-1} is produced. Under optimum conditions this population level could be attained within 4 to 9 days for *Rhizobium*; 5 to 7 days for *Azospirillum*; 2 to 3 days for phosphobacteria and 6-7 days for *Azotobacter*.

Inoculum preparation

- Prepare appropriate broth for specific to the bacterial inoculant in 250 ml, 500 ml, 3 litre and 5 litre conical flasks and sterilize.
- The broth in 250 ml flask is inoculated with efficient bacterial strain under aseptic condition.
- Keep the flask under room temperature in rotary shaker (200 rpm) for 5-7 days.

- Observe the flask for growth of the culture and estimate the population, which serves as the starter culture.
- Using the starter culture (at log phase) inoculate the larger flasks (500 ml, 3 liters and 5 liters) containing the broth, after obtaining growth in each flask.
- The above media is prepared in large quantities in fermentor

Fermentor

Mass culturing requires large capacity fermentors. Fermentor is the vessel which maintains the controlled environmental conditions for the growth of microorganisms and provides access for inoculation, sampling, aeration and cleaning. It should be made of stainless steel to withstand high pressure and also to resist corrosion. High quality fermentor will have smooth surfaces inside.

Basic functions of a fermentor

The main function of a fermentor is to provide a controlled environment for the growth of microorganisms in order to obtain a desired product.Fermentor vessel should be capable of being operated aseptically for a number of days. It should withstand high pressure. The power consumption should be as low as possible. The vessel should be designed to require minimal use of labour for operation.The vessel should be constructed to ensure smooth internal surfaces without cracks and crevices.

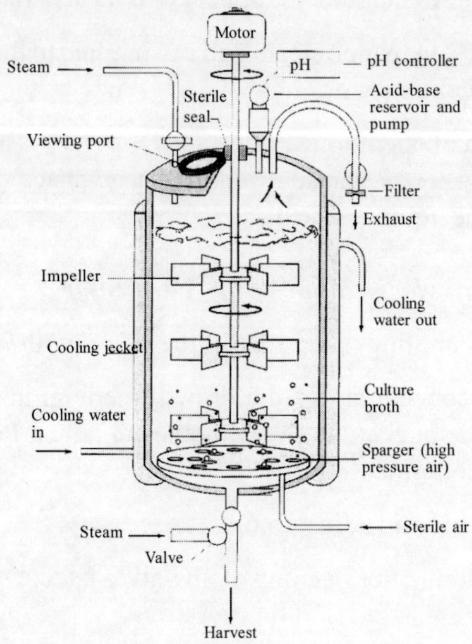

Description of a fermentor

The vessel in which industrial fermentation process carried out is called as fermentor. The volume of the fermentor varies from 5-10 lit to 500,000 lit. Similarly, the fermentor model will vary for aerobic process and for anaerobic process. It is domed at the top and bottom and held upright by a stainless steel skirt. The top has accessories like a steam pressure gauge, a safety pressure release valve, aeration system with filters, inoculation port. Inlet and outlet ports are provided for water passage through stainless steel cooling coil. The sampling port is provided at the bottom to withdraw samples or to harvest the cultures. The basic structure of fermentor is as follows:

The components are,

Inlet - to feed the substrates, microorganisms etc.

Outlet - to collect the products, to take samples.

Stirrer - to mix the contents so as to get contact between nutrients and microbes.

Air sparger - to provide oxygen to the microbes by small air bubbles.

pH meter/controller - to check the pH and if needed to add acid and alkali to maintain optimum pH.

Thermometer - to check the temperature.

Outer jacket - by which cool water will be circulated to cool the fermentor (because during fermentation, lot of heat will be generated).

Sterilization unit - by which steam will be supplied to the fermentor to sterilize the medium and container.

However, the actual operation volume in a fermentor is always less than that of the total volume (2/3), because it requires head space for splashing, foaming and aeration of the liquid inside.

Sterilizing the growth medium in the Fermentor

- Close the sampling valve and fill the vessel with 90 litre of clean water.
- Prepare a concentrate of the growth medium in 10 litre of water in a clean plastic bucket or other suitable container. Pour into the 90 litre of water in the fermentor.
- Check the pH and adjust to 6.8, if necessary.
- Close the fermentor opening of all valves except the air outlet valve.

- Bring the growth medium to boiling under maximum heating. Allow the steam to escape.
- Close the air outlet valve and allow pressure to buildup inside the vessel. Maintain a pressure of 15 lb/in^2 at 121°C for 20 min.
- Switch off the fermentor and cool the medium by circulating cool water.

Mass culturing in fermentor

- Spray the inoculation port with alcohol and flame thorough-ly. Allow the port to cool.
- Allow the starter culture to flow into the steel fermentor. (inoculum from seed tank fermentor/ large flasks is added at the rate of 1% by volume)
- Open the air outlet valve, turn on the air pump.
- Regulate the air flow to 3-10 litre of air per hour per litre of medium. The sterile air provides aeration as well as agitation for the culture growth.
- Draw samples and analyze for growth, periodically if necessary.
- Once the culture reaches full growth, turn off the air supply and the cells are harvested with the population load of 10^9 cells ml^{-1} after incubation period through the sampling port.
- There should not be any fungal or any other bacterial contamination at 10^{-6} dilution level.
- It is not advisable to store the broth after fermentation for periods longer than 24 hours. Even at 4°C number of viable cells begins to decrease.

The incubation period in the fermentor can be reduced by increasing the inoculum level upto 5%. Aeration and agitation of cultures hasten the growth. Cultures evolve heat while metabolically active; so fermentor must be cooled by passing water through outer jacket to maintain optimum temperature. The bacterial strain, initial inoculum level, aeration, temperature and incubation period influences the output of cells. The prime objective is to attain high population in minimum time.

2. Processing of carrier material

The use of ideal carrier material is necessary in the production of good quality biofertilizer. Peat soil, lignite, vermiculite, charcoal, press mud, farmyard manure and soil mixture can be used as carrier materials. The neutralized peat soil/ lignite are found to be better carrier materials for biofertilizer production The following points are to be considered in the selection of ideal carrier material.

- Cheaper in cost.
- Should be locally available.
- High organic matter content.
- No toxic chemicals.
- Water holding capacity of more than 50%.
- Easy to process, friability and vulnerability.

The following steps are followed in the processing of the carrier material.

- The carrier material (peat or lignite) is powdered to a fine powder so as to pass through 212 micron IS sieve.
- The pH of the carrier material is neutralized with the help of calcium carbonate (1:10 ratio), since the peat soil / lignite are acidic in nature (pH 4-5).
- The neutralized carrier material is sterilized in an autoclave to eliminate the contaminants.

3. Mixing the carrier and the broth culture

Inoculant packets are prepared by mixing the broth culture obtained from fermentor with sterile carrier material as described below :

Preparation of Inoculants Packet

The neutralized, sterilized carrier material is spread in a clean, dry, sterile metallic or plastic tray.The bacterial culture drawn from the fermentor is added to the sterilized carrier and mixed well by manually (by wearing sterile gloves) or by mechanical mixer. The culture suspension is to be added to a level of 35 - 40% water holding capacity depending upon the population.The inoculant packet of 200 g quantities in polythene bags, sealed with electric sealer and allowed for curing for 2-3 days at room temperature (curing can be done by spreading the inoculant on a clean floor/polythene sheet/ by keeping in open shallow tubs/ trays with polythene covering for 2-3 days at room temperature before packaging).

Specification of polythene bag

- The polythene bags should be of low density grade.
- The thickness of the bag should be around 50-75 micron.

- Each packet should be marked with the name of the product, strain number, the crop to which recommended, method of inoculation, date of manufacture, batch no., date of expiry, price, full address of the manufacturer and storage instruction etc.

Storage

At normal room temperature the inoculant can be stored up to 6 months. The shelf life could be extended to 12 months when stored at 4°C. The inoculant packets should be stored in a cool place away from heat or direct sun light. The packets may be stored at room temperature or in cold storage conditions in lots in plastic crates or polythene / gunny bags. The population of the organisms in the inoculant packet may be determined at 15 days interval. The population should be more than 10^8 cells / g for *Rhizobium*, 10^9 cells / g for *Azospirillum* and 10^7 cells / g of the carrier material for *Azotobacter* on dry weight basis immediately after curing and should not be less than 10^7 cells / g for *Rhizobium* and *Azospirillum* and 10^6 cells / g of carrier material on dry weight basis 15 days before the expiry date.

Flow diagram of production

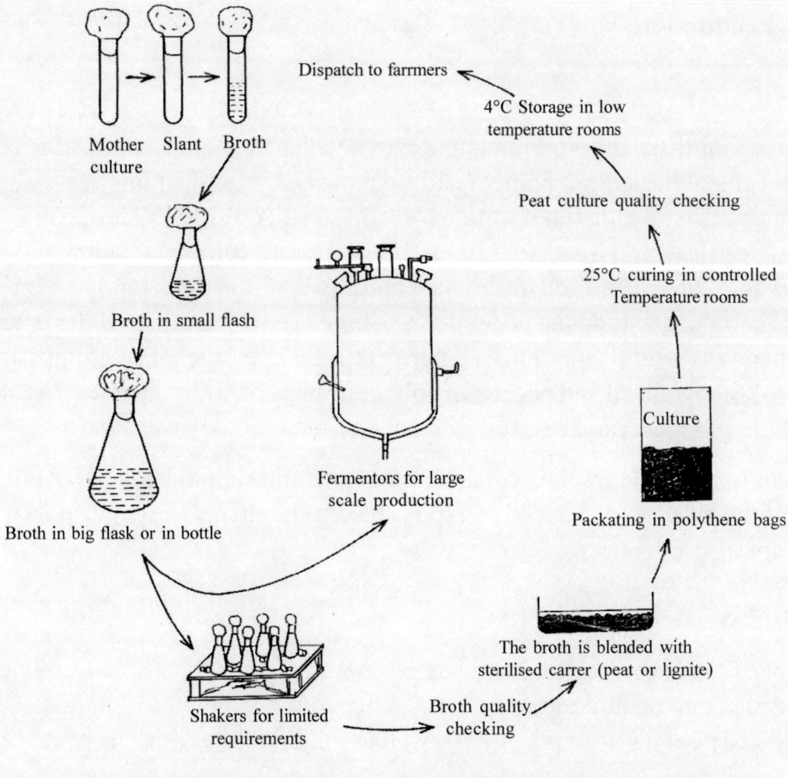

Mass Production of Fungal Inoculants (*Trichoderma viride*)

Dual Culture Technique

This technique is used to test the efficacy of *T. viride* against the target pathogen. In a sterile petri plate 15 ml of sterilized and cooled potato dextrose agar (PDA) was poured. To one end of the petri plate, 8mm disc of *T. viride* was placed aseptically and in the opposite end, 8mm disc of the pathogen culture was placed. The plates were incubated at room temperature and periodically observed for the antagonistic activities like physical parasitism, overgrowth on pathogen, production of lytic enzymes, inhibition zone etc., to screen the bioefficacy of antagonists.

Preparation of Nucleus Culture

Individual colonies of *Trichoderma viride* grown on TSM were transferred aseptically to Potato dextrose agar medium following single hyphal tip method or single spore isolation method. The plates are incubated at room temperature for 4 to 5 days and it serves as the nucleus culture. Simultaneously, slant cultures were also prepared and maintained for future use.

Mass Production

i) Small Scale

The mycelial discs grown in potato dextrose agar medium was transferred to 250 ml conical flasks containing sterilized molasses yeast mediun (Yeast extract - 5g; molasses - 30g, distilled water -1000 ml sterilized at 121°C for 20 min). @ 100 ml per flask and incubated for 10 days. The mycelial mat along with the broth was homogenized and mixed with talc powder (500 mesh) @ 1:2. Carboxy methyl cellulose was added @ 5g per kg of the product as sticker. The product was shade dried until it attains 12 per cent moisture level. The clumps were broken, homogenized and packed in polythene bags of 100g capacity. The shelf life of the product was 120 days.

The required population level at the time of packing should be 28 x 10^7 cfu per gram of the product. At the end of expiry the cfu should not be less than 20 x 10^6 per gram of the product.

ii) Large Scale

Multiply *T. viride* on potato dextrose agar in 9 cm diameter petriplates. Melt the sterilized potato dextrose agar in flasks. After cooling, transfer 15ml of medium to sterilized petriplates aseptically. Allow it to solidify. Take with help of sterilized inoculation needle, a small portion of the *T. viride* culture from agar slants and place at the centre of petridish. Incubate for 5 days at room temperature. This culture is used for subsequent inoculations.

Distribute 70 ml of molasses yeast medium in each of 250ml conical flasks and plug with non-absorbent cotton wool. Autoclave at 15 lbs/sq. inch pressure (121°C) for 20 – 30 min. Take 8 mm mycelial disc of *T. viride* from the petridish and transfer to each flasks. Incubate for 10 days. This mother culture is used for inoculating medium in fermentor.

Prepare molasses-yeast medium (50 litre.) in 75 lt. fermentor and sterilize at (121°C) 15 lb/sq. inch. and cool to room temperature. Add 2.5 1 of inoculum (mother culture) grown in conical flasks aseptically into the sterilized medium Incubate for 10 days. Draw samples at 10 day after incubation and assess the spore load in the broth. The spore load should be at least 10^8 spores / ml at the time of sampling. Provide aeration in the fermentor for 4-8 hr/day by agitation. Presence of greenish fungal growth indicates *T. viride.* Agitate the entire medium and drain the contents. This broth culture contains fungal mat, spores and antibiotics.

Avoiding contamination

The fermentor should be located in a dust free room. Mixing of the broth with the talc powder, drying and packing should be done in a dust free room. Packing should be done only after a thorough drying to 20% moisture content.

Processing

Mix 1 liter for broth culture with 2 kg talc powder. Talc powder/French chalk powder/Soapstone powder (Super white). Air dry and add carboxy methyl cellulose @ 5g/kg.

Formulation

Talc based *Trichoderma viride* formulation is ready for use after air drying and sieving (moisture content 12%). It is packed in opaque polythene bags and stored in a cool dry place before use. The product can be safely stored for four months without loosing its viability.

Mass Production of *Azolla*

Azolla is a free-floating water fern that floats in water and fixes atmospheric nitrogen in association with nitrogen fixing blue green alga *Anabaena azollae*. Rice crop response studies with *Azolla* biofertilizer in India, China and Vietnam have provided good evidence that *Azolla* incorporation into the soil as a green manure crop is one of the most effective ways of providing nitrogen source for rice. The utilization of *Azolla* as dual crop with wetland rice is gaining importance in Philippines, Thailand, Srilanka and India. The important factor in using *Azolla* as a biofertilizer for rice crop is its quick decomposition in soil and efficient availability of its nitrogen to rice. In tropical rice soils the applied *Azolla* mineralizes rapidly and its nitrogen is available to the rice crop in very short period. The common species of *Azolla* are *A. microphylla, A. filiculoides, A. pinnata, A. caroliniana, A. nilotica, A. rubra and A. mexicana.*

A simple nursery method for large scale multiplication of *Azolla* in the field has been evolved for easy adoption by the farmers.

i) Select a wetland field and prepare thoroughly and level uniformly.

ii) Mark the field into one cent plots (20 x 2m) by providing suitable bunds and irrigation channels.

iii) Maintain water level to a height of 10 cm.

iv) Mix 10 kg of cattle dung in 20 litres of water and sprinkle in the field.

v) Apply 100 g superphosphate as basal dose.

vi) Inoculate fresh *Azolla* @ 8 kg to each pot.

vii) Apply superphosphate @ 100 g as top dressing fertilizer on 4th and 8th day after *Azolla* inoculation.

viii) Apply furadan granules @ 100 g/plot on 7th day after *Azolla* inoculation.

ix) Maintain the water level at 10 cm height throughout the growth period of two or three weeks.

Within a period of three weeks, the entire plot is covered with *Azolla* like a green mat. Harvest and collect the *Azolla* biomass by draining the water in the plot.

Mass production of *Azolla* spore inoculum

Azolla sporocarp inoculum may be produced by the following method.

i) Prepare the field well and level it uniformly and mark the plots each with a size of 40 m^2 area and form raised bunds around the plot.

ii) Prepare 10 kg fresh cattle dung mixture in 20 l of water and sprinkle over the plot.

iii) Inoculate the fresh young fronds of *Azolla* @ 8 kg per plot.

iv) Apply 100 g super phosphate in three split doses at 4 days interval and furadan granules at 100 g per plot.

v) Allow the *Azolla* cultures to grow for 4-5 weeks period.

vi) *Azolla* cultures turn into yellowish colour which indicates the spore production. Examine the back side of the fronds for the occurrence of spores.

vii) Allow the culture for another 1-2 weeks to get the matured sporocarps.

viii) Harvest the sporulated fronds and make into heaps and cover with a layer of clay soil.

ix) Moist the heaps periodically by sprinkling water and allow them to decompose for 3 - 4 weeks.

x) Spread the decomposed material and dry it under shade.

xi) Store the dried frond-based spore inoculum in gunny bags and use it for inoculation into rice fields.

Examine the dried spore inoculum under microscope and note the presence of megasporocarps and microsporangia.

Mass Production of Cyanobacteria (BGA)

Blue green algal inoculation with composite cultures was found to be more effective than single culture inoculation. A technology for mass scale production of composite culture of blue green algae under rice field condition was developed at Tamil Nadu Agricultural University (TNAU) and the soil based BGA inoculum could survive for more than 2 years. At many sites where algal inoculation was used for three to four consecutive cropping seasons, the inoculated algae establish well and the effect persisted over subsequent rice crop.

The blue green algal inoculum may be produced by several methods *viz*., in tubs, galvanized trays, small pits and also in field conditions. However the large-scale production is advisable under field condition which is easily adopted by farmers.

a. Multiplication in trays

a. Big metallic trays (6'x 3'x 6"lbh) can be used for small scale production

b. Take 10 kg of paddy field soil, dry powder well and spread

c. Fill water to a height of 3"

d. Add 250 g of dried algal flakes (soil based) as inoculum

e. Add 150 g of super phosphate and 30 g of lime and mix well with the soil

f. Sprinkle 25 g Carbofuran to control the insects

g. Maintain water level in trays

h. After 10 to 15 days, the blooms of BGA will start floating on the water sources

i. At this stage stop watering and drain. Let the soil to dry completely

j. Collect the dry soil based inoculum as flakes

k. Store in a dry place. By this method 5 to 7 kg of soil based inoculum can be obtained.

b. Multiplication under field condition

i. Select an area of 40 m^2 (20m x 2m) near a water source which is directly exposed to sunlight.

ii. Make a bund all around the plot to a height of 15 cm and give it a coating with mud to prevent loss of water due to percolation.

iii. Plot is well prepared and levelled uniformly and water is allowed to a depth of 5-7.5 cm and left to settle for 12 hrs.

iv. Apply 2 kg of superphosphate and 200 g lime to each plot uniformly over the area.

v. The soil based composite starter culture of BGA containing 8-10 species @ 5 kg / plot is powdered well and broadcasted.

vi. Carbofuran @ 200 g is also applied to control soil insects occurring in BGA.

vii. Water is let in at periodic intervals so that the height of water level is always maintained at 5 cm.

viii. After 15 days of inoculation, the plots are allowed to dry up in the sun and the algal flakes are collected and stored.

The floating algal flasks are blue green in colour. From each harvest, 30 to 40 kg of dry algal flakes is obtained from the plot.

Mass Production of Mycorrhizal fungi

The commercial utilization of mycorrhizal fungi has become difficult because of the obligate symbiotic nature and difficulty in culturing on laboratory media. Production of AM inoculum has evolved from the original use of infested field soils to the current practice of using pot culture inoculum derived from the surface disinfected spores of single AM fungus on a host plant grown in sterilized culture medium. Several researches in different parts of the world resulted in different methods of production of AM fungal inoculum as soil based culture as well as carrier based inoculum. Root organ culture, nutrient film technique and aerophonic culture provide scope for the production of soil less culture.

Carrier based AM inoculum

As a carrier based inoculum, pot culture is widely adopted method for production. The AM inoculum was prepared by using sterilized soil and wide array of host crops were used as host. As the sterilization process is a cumbersome one, scientists started using inert materials for production of AM fungi. The researchers tried use of perlite, montmorillonite clay etc., Use of Vermiculite as substrate for the replacement of soil sterilization was tested at Tamil Nadu Agricultural University, Coimbatore by using maize as host crop. The best inoculum was obtained in *Glomus fasciculatum* which recorded 90 per cent infection in maize roots on 45 days.

The mass production technique is as follows :

- Construct a brick tank with 1.0 m (B) x 2.0 m (L) x 0.30 m (H).
- Two layers of brick is sufficient for bottom layer - There is no need to plaster on both sides of tank
- Fill up 150 kg of Vermiculite upto 20 cm height of the tank
- Add 200 g of good pure AM fungal inoculum + add 2g of urea + 4 g of super phosphate at the time of sowing seeds.
- Sow the surface sterilized seed of Maize (50g).
- Maintain the maize plant for 45 days and collect the roots and note the infection percentage.
- Grow the stock plants for 60 days (8 weeks) and cut the upper parts of the plant.
- Cut all the roots of the plant and mix thoroughly with vermiculite in the trench.
- The inoculum produced consists of a mixture of vermiculite, spores, pieces of hyphae and infected root pieces.

- Thus within 60 days, 55 kg of AM inoculum could be produced from 1 sq meter area.

The final product should be stored in gunny bags. Composted coir pith could also be used as substrate for large scale production of AM inoculum. This inoculum will be sufficient to treat 550 m^2 nursery area having 11,000 seedlings.The shelf-life of AM inoculum is six months.

30

Quality Control of Biofertilizers

Biofertilizers are one of the important components of organic farming and sustainable agriculture. Awareness among farmers is increasing on the biological fertilizers. But, the success of biofertilizer inoculation depends on the quality of the inoculant used. Because farmers are unable to judge the quality of the inoculants at the time of purchase or even at use. The inoculant producers are negligent to institute quality control programmes. Unfortunately much of the inoculant produced in the world today is relatively of poor quality. In such situation maintaining high quality inoculant products depend on the internal implementation of effective quality control systems by manufacturer. A major goal of a quality standard is to control the quality of inoculant to be used in any country.

International Perspectives on Quality Control

There are wide variations in the quality control programmes and parameters in developing countries and developed countries. The following are the perspectives on the quality control of microbial inoculants at international levels:

- Regulation on inoculant quality varies from country to country and no set of international standards exists.
- Brazil, Canada, France and Uruguay have regulatory authorities supported by legislation.
- Australia, South Africa and New Zealand have quality control programmes in which inoculant manufacturers participate voluntarily.
- In many countries (including USA, UK, India) product quality standards are left to the discretion of the manufacturers.
- A great disparity exists throughout the world today in the price, quality and effectiveness of microbial inoculants available to farmers.

France currently has the highest standards world wide for inoculant quality, and also has a strict requirement that no contaminants be present in rhizobial inoculant products. Australian standards are also high, requiring 1 x 10^9 rhizobia per g moist inoculant and fewer than 10^6 contaminants per g. England has no governmental standard, but Agricultural Genetics Company's self imposed minimum is 2 x 10^9 rhizobia per g with a maximum of 10^6 contaminants per g. The Thai standard for its government programme which produces inoculant is 5 x 10^7 rhizobia per g, but this minimum is sensibly tied to a requirement for a high inoculant application rate with the net result that 10^4-10^6 *Bradyrhizobium japonicum* per soybean seed are applied.

The Netherlands require from 4 to 25 x 10^9 rhizobia per g, and South Africa and New Zealand both require a minimum of 1 x 10^8 rhizobia per g. Rwanda prescribes 1 x 10^9 rhizobia per g and the contaminants should not be more than 0.001% of the rhizobial count. The United States applies no standards and inoculant quality is left to the manufacturer's discretion, but most US inoculants exported to Canada contain at least 1 x 10^8 rhizobia per g. The Canadian standard for minimum numbers of rhizobia in inoculants therefore varies with manufacturer, legume host, and application rate. For inoculants tested in Canada in 1992, the required number varied from 2.7 x 10^7 to 1.6 x 10^9 rhizobia per g.

Quality control must begin with the maintenance of mother culture and then in broth culture before addition to the carrier and also for the finished product.The following points have to be taken into consideration while adopting quality control procedures for biofertilizers with the objectives of ensuring supply of quality inoculants to the farmers.

1. Biofertilizer production unit should have adequate microbiological laboratory facilities to produce quality inoculum. The unit should be maintained by a qualified microbiologist. The production unit should have one quality control laboratory to check their own biofertilizer packets.

2. Each biofertilizer production unit should maintain crop specific and pretested efficient N_2 fixing microorganisms or phosphobacteria either collected from a reputed scientific organization or of their own. The maintenance of microorganisms should be as per the standard microbiological procedures and repeated sub-culturing should be avoided. If possible, lyophilized cultures should be used for each batch of biofertilizer production. Manufacturer should indicate the source of the culture in the biofertilizers packet produced.

3. A biofertilizer production unit should use fermentors so as to ensure high microbiological counts at broth level itself.

4. The biofertilizer production unit should use good quality carrier material like Peat, lignite, wood charcoal or any other suitable carrier and the biofertilizer products should indicate the name of the carrier.

5. The inoculum packet should contain the exact quantity of biofertilizer, which is mentioned in the packet.
6. The microbial inoculum must contain the specified population count at the time of packing and expiry. The packet should be crop specific and should give effective yield increase. The production centers and distribution centers should store the biofertilizer packets in cool temperature so as to ensure longer shelf life.

Suggestions for Distribution Agencies

1. Keep biofertilizer packets away from direct heat and sunlight. Store it in cool and dry place.
2. Sell only biofertilizer packets which contain batch number, the name of the crop, the date of manufacture and expiry period.
3. If the expiry period of the packet is over, discard it, it will not be effective.
4. Keep biofertilizer packets away from fertilizers or pesticide containers.
5. Don't prick holes into the packets or puncture them.
6. Don't store heavy items above biofertilizer packets.

A. *Rhizobium*

1. Control of the broth culture

To ensure that the culture broth is free from contamination and also the population of rhizobial cells, the broth culture can be streaked on glucose peptone agar and lactose agar. The slow growth or the absence of growth in the latter ensures that the inoculant is free from *Agrobacterium*, a common contamination with *Rhizobium*. A smear of the broth culture can be stained as per gram staining and the absence of gram-positive cells indicate that the inoculant is free from contaminants.

2. Control of the finished inoculant

The number of rhizobia cells is ascertained by serial dilution plate technique with yeast extract mannitol medium. One gm of inoculant is serially diluted with sterile distilled water upto 10^8dilution and plated. After incubation period the rhizobial colonies are identified by their nature of growth. The presence of contaminating bacteria could also be identified (The normal contaminant *Agrobacterium* takes the red colour of the congored where as *Rhizobium* does not).

3. Plant infection test

This test is for the authentication of rhizobia. Good quality leguminous seeds preferably of the respective host are to be used. The seeds are surface sterilized with 80% ethanol and immersed for 3 minutes in acidified 1% (W/V) solution of mercuric chloride. Then the seeds washed thoroughly with sterile distilled water at least for five times. The surface sterilized seeds are pre-germinated over soft agar (1%) plates. In glass tube of 30x 4 cm size germination paper is inserted giving provision for holding the seeds in the topfold of the paper. This set up is sterilized in an autoclave. The pregerminated seeds are aseptically transferred (3 seeds/tube) to the topfold in such a way that the root portions are in the bottom side. Well-grown culture of *Rhizobium* is poured at the rate of one ml on each seed. Nitrogen free nutrient solution of about 25 ml per tube is initially added and 10-15 ml at weekly intervals.

After thirty days of sowing the plants are removed along with the paper from the tube. The root system should have bold pink coloured nodules.

BIS specifications for *Rhizobium* inoculant

Rhizobium inoculant should contain a minimum of 10^8 viable *Rhizobium* cells/g of the carrier on dry weight basis at the time of manufacture and 10^7 before the expiry date marked on the packet (25-30° C storage temperature).

- *Rhizobium* inoculant should have no contamination with other microorganisms at 10^6 dilution.
- The pH of inoculant should be 6.0 to 7.5.
- Carrier material should be neutral and pass through 100-micron sieve.
- Should be packed in 50-75 micron low-density polythene bag.
- Each packet should be marked with the following particulars – Name of the product, leguminous crop for which intended, date of manufacture, date of expiry and storage instructions.

B. *Azospirillum*

The *Azospirillum* inoculant packets may be subjected to the quality control tests during preparation as well as during storage. The quality of the *Azospirillum* inoculant may be tested by the following methods.

1. Determination of *Azospirillum* population in inoculant by MPN technique

2. Checking the presence of contaminating bacteria in the inoculant by serial dilution technique using nutrient agar medium.

Quality control specifications developed by TNAU for *Asozpirillum* inoculant.

- *Azospirillum* inoculant should contain a minimum of 10^9 viable *Asozpirillum* cells / g of dry carrier at the time of manufacture and 10^7 cells / g dry carrier at 15 days before the expiry date mentioned on the packets.
- No contamination with other microorganisms.
- pH of inoculant should be 7.2-7.5.
- Carrier material should be passed through 100 micron IS sieve.
- Each packet should be marked with strain number, date of manufacture, date of expiry, method of application, etc.

C. Phophate Solubilizing Microrrganisms (Phosphobacteria)

The following specifications should be adopted by entrepreneurs to have a quality phosphobacteria inoculant.

1. The peat based phosphobacteria inoculant should be contain a minimum of 10^9 viable cells / g of peat based inoculant at the time of manufacture and 10^7 viable cells / g before the expiry marked on the packet.
2. The pH of the peat based inoculant should be between 6.5 and 7.5.
3. The carrier material used for phosphobacteria inoculant preparation should not contain toxic material and should satisfy the required characters of high water holding capacity and high organic carbon content.
4. The carrier material prepared should pass through 100-micron sieve.
5. The polythene bag used for packing phosphobacteria should be of low density (50-75μ).
6. The front page of polythene bag should be printed with the name of biofertilizer, crops for which the inoculant to be used, application methods, methods of storage, date of manufacture, date of expiry, net weight and price.
7. Phosphobacteria inoculant should have no contamination with other microorganisms at 10^6 dilutions.

Bureau of Indian Standards (BIS) for carrier- based Biofertilizers

BIS have notified specifications for *Azotobacter* (ISI, 1979) and *Rhizobium* (ISI, 1986) long back for use of biofertilizer producers. The committee has also suggested specifications for phosphate solubilizing microorganisms and *Azospirillum*.

Parameters	*Rhizobium*	*Azotobacter*	*Azospirillum*	PSM
Base	Carrier based	Carrier based	Carrier based	Carrier based
Cell No. at the time of manufacture	10^7	10^6	10^7	10^8
Contamination	Nil at 10^5 dilution	Nil at 10^5 dilution	Nil at 10^5 dilution	Nil at 10^6 dilution
Expiry period from date of manufacture	6 months	6 months	6 months	6months
pH	6.5 – 7.5	6.5 - 7.5	6.5 - 7.5	6.5 - 7.5
Strain	Should be checked serologically	Nothing specific *A. chroococcum* sp. is mentioned	Nothing specific	Nothing specific
Carrier should pass through	150-212 micron IS sieve	106 micron IS sieve	106 micron IS sieve	106 micron IS sieve
Others	Nodulation test + ve result in 50% or more dry matter than control	Minimum amount of N-fixation not less than 5mg/g of sucrose utilized	Should produce white pellicle in NFB media at 10^7 dilution	Should show phosphate solubilizing zones of 12 mm in tricalcium phosphate medium at 10^5 dilution

31

Method of Application of Biofertilizers

Bacterial biofertilizers are supplied as carrier based inoculants. Peat or lignite is used as carrier material. Carrier based bacterial inoculants are applied by the following methods.

i. Seed treatment or seed inoculation.

ii. Seedling root dip and

iii. Main field application

Seed Treatment

One packet of the inoculant (200 g) 7is mixed with 200 ml of rice kanji to make a slurry. The seeds required for an acre are mixed in the slurry so as to have a uniform coating of the inoculant over the seeds and then shade dried for 30 minutes. The shade dried seeds should be sown within 24 hours. One packet of the inoculant (200 g) is sufficient to treat 10 kg of seeds.

Seedling Root Dip

This method is used for transplanted crops. Two packets of the inoculant is mixed in 40 litres of water. The root portion of the seedlings required for an acre is dipped in the mixture for 5 to 10 minutes and then transplanted.

Main Field Application

Four packets of the inoculant is mixed with 20 kgs of dried powdered farm yard manure and then broadcasted in one acre of main field just before transplanting.

Rhizobium

For all legumes *Rhizobium* is applied as seed inoculant. While recommending *Rhizobium* biofertilizer, cross inoculation grouping (CIG) should be taken care.

Asozpirillum / Azotobacter

In the transplanted crops, *Asozpirillum* is inoculated through seed, seedling root dip and soil application methods. For direct sown crops, *Asozpirillum* is applied through seed treatment and soil application.

Phosphobacteria

Inoculated through seed, seedling root dip and soil application method as in the case of *Asozpirillum*.

Combined application of bacterial biofertilizers

Phosphobacteria can be mixed with *Asozpirillum* and *Rhizobium*. The inoculants should be mixed in equal quantities and applied as mentioned above.

Points to remember

- Bacterial inoculants should not be mixed with insecticide, fungicide, herbicide and fertilizers.
- When seeds are to be treated with any chemical, seed treatment with bacterial inoculant should be done last, prior to sowing.
- Quantity of biofertilizers requirement for different crops is furnished in Appendix 1.

Appendix - I

S.	Crop	Seed	Nursery	Seedling dip	Main Field	Total requirement of packets
Azospirillum						
1.	Rice	5	10	5	10	30
2.	Sorghum	3	-	-	10	13
3.	Pearl millet	3	-	-	10	13
4.	Ragi	3	-	5	10	18
5.	Maize	3	-	-	10	13
6.	Sunflower	3	-	-	10	13
7.	Castor	3	-	-	10	13
8.	Sugarcane	10	-	-	36 (3 split at 30 days interval)	46
9.	Turmeric	-	-	-	24 (2 split)	24
10.	Tobacco	1	3	-	10	14
11.	Papaya	2	-	-	10 g/pit	-
12.	Mandarin orange	2	-	-	10 g/pit	-

contd...

13.	Tomato	1	3	-	10	14
14.	Banana	-	-	5	10 g/pit	-
15.	Cotton	3	-	-	10	13
Rhizobium						
1.	Soybean	5	-	-	-	5
2.	Groundnut	5	-	-	-	5
3.	Bengal gram	5	-	-	-	5
4.	Black gram	3	-	-	-	3
5.	Green gram	3	-	-	-	3
6.	Red gram	3	-	-	-	3
7.	Cowpea	3	-	-	-	3

Method of application of *Azolla* to rice crop

The *Azolla* biofertilizer may be applied in two ways for the wetland paddy. In the first method, fresh *Azolla* biomass is inoculated in the paddy field before transplanting and incorporated as green manure. This method requires huge quantity of fresh *Azolla*. In the other method, *Azolla* may be inoculated after transplanting of rice and grown as dual culture with rice and incorporated subsequently.

a. *Azolla* biomass incorporation as green manure for rice crop

- Collect the fresh *Azolla* biomass from the *Azolla* nursery plot.
- Prepare the wetland well and maintain water just enough for easy incorporation.
- Apply fresh *Azolla* biomass (15 t ha^{-1}) to the main field and incorporate the *Azolla* by using implements or tractor.

b. *Azolla* inoculation as dual crop for rice

- Select a transplanted rice field.
- Collect fresh *Azolla* inoculum from *Azolla* nursery.
- Broadcast the fresh *Azolla* in the transplanted rice field on 7th day after planting (500 kg / ha).
- Maintain water level at 5-7.5cm.
- Note the growth of *Azolla* mat four weeks after transplanting and incorporate the *Azolla* biomass by using implements or tractor or during inter-cultivation practices like weeding.

- A second bloom of *Azolla* will develop 8 weeks after transplanting which may be incorporated again during second weeding.
- By the two incorporations, 20-25 tonnes of *Azolla* can be incorporated in one hectare rice field.

Method of Application of Cyanobacteria (BGA) in rice field

Blue green algae may be applied as soil based inoculum to the rice field following the method described below.

a. Powder the soil based algal flakes very well.

b. Mix it with 10 Kg soil or sand (10kg powdered algal flakes with 10 Kg soil / sand).

c. BGA is to be inoculated on 7-10 days after rice transplanting.

d. Water level at 3-4" is to be maintained at the time of BGA inoculation and then for a month so as to have maximum BGA development.

e. A week after BGA inoculation, algal growth can be seen and algal mat will float on the water after 2-3 weeks. The algal mat colour will be green or brown or yellowish green.

Method of Application of AM Fungi

a. For Nursery application:

100 g bulk inoculum is sufficient for one-meter square. The inoculum should be applied at 2-3 cm below the soil at the time sowing. The seeds/cuttings should be sown/planted the AM inoculum and cause infection.

b. For polythene bag raised crops *(Forest trees, coffee & tea etc.)*

50 to 100 g of bulk inoculum is sufficient for each packet. Mix 10 kg of inoculum with 1000kg of sand potting mixtures and pack the potting mixtures in polythene bag before sowing.

c. AM inoculum required for out planting

Twenty grams of AM inoculum is required for planting per seedling. Apply inoculum at the time of planting.

d. AM inoculum required for existing trees

Two hundred gram of AM inoculum is required for inoculating one tree, apply inoculum near the root surface at the time of fertilizer application.

32

Biofuel Production

At this moment, the majority of energy is obtained from fossil fuels and requires switching over to alternative energy sources is the need of the hour. Fossil fuels are nonrenewable energy source and also have seriously negative impacts on the environment. The use of fossil fuels cause excessive global climate change because emissions of greenhouse pollutants and the formation of compounds CO_x, NO_x, SO_x, C_xH_y, ash and other organic compounds that are released into the atmosphere. The increase in greenhouse gas emission will result in global warming, climate change, environmental degradation and health problems.

Biofuel, any fuel that is derived from biomass - that is, plant or algae material or animal waste. Since such feedstock material can be replenished readily, biofuel is considered to be a source of renewable energy, unlike fossil fuels such as petroleum, coal, and natural gas.

First Generation Biofuels

The first generation biofuels refer to the fuels that have been derived from sources like starch, sugar, animal fats and vegetable oil. The oil is obtained using the conventional techniques of production.

Second-Generation Biofuels

They are also known as advanced biofuels, that can be manufactured from various types of biomass (cellulose). Biomass is a wide-ranging term meaning any source of organic carbon that is renewed rapidly as part of the carbon cycle.

Third Generation Biofuels

It is based on improvements in the production of biomass. It takes advantage of specially engineered energy crops such as algae as its energy source. The algae are cultured to act as a low-cost, high-energy and entirely renewable feedstock.

Bio Hydrogen Production

The classical methods of producing hydrogen include steam reforming of natural gases, coal gasification and electrolysis of water. Conventional hydrogen gas production methods are energy intensive processes requiring high temperatures (>840°C) and not environmental friendly. Bio-hydrogen is ideal as it can be operated at ambient temperature and pressure with minimal energy consumption and are more environmental friendly. The major biological processes for bio-hydrogen production are bio-photolysis of water by algae, dark fermentation, photo-fermentation of organic materials and the sequential dark and photo-fermentation processes. Biohydrogen production methods can be broadly categorized into four primary groups as shown in figure.

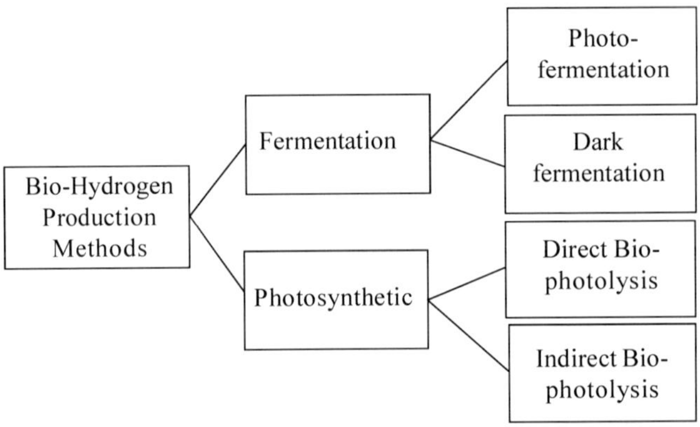

Different microorganisms that have been studied for bio-hydrogen production such as green algae, cyanobacteria (blue-green algae), photosynthetic bacteria and fermentative bacteria, archaea, anaerobic and facultative aerobic bacteria, and lower eukaryotes (i.e., green algae and protists).

Direct biophotolysis

Biophotolysis is the action of light on biological systems that result in dissociation of water into molecular hydrogen and oxygen; $H_2O \rightarrow H_2 + \frac{1}{2} O_2$. The solar energy is directly converted to hydrogen. $2H_2O$+Light energy $\rightarrow 2H_2+O_2$. Cyanobacteria or green microalgae, can use light to carry out photosynthesis (they possess· chlorophyll a and the photosynthetic systems: PSII and PSI). The pigments in PSII (P680) absorb the photons with a wavelength shorter than 680 nm, generating a strong oxidant capable of splitting water into protons (H^+), electrons (e^-) and O_2. The electrons reduce the ferredoxin (Fd) and/or nicotinamide adenine dinucleotide phosphate ($NADP^+$) into their reduced forms. Under special conditions, the reduced ferredoxin can also be used by hydrogenase or nitrogenase to reduce protons for evolution of molecular hydrogen ($2H^+ + 2Fd \rightarrow H_2 + 2Fd$).

Indirect biophotolysis

Indirect biophotolysis avoids the inhabitation of hydrogenase by separating the hydrogen production process from the oxygen production process into two stages. At first it involves the splitting of water molecules by sunlight to produce protons and oxygen and at the same time carbon dioxide fixation occurs to produce storage carbohydrate, followed by the production of hydrogen gas by hydrogenase:

$$12H_2O + 6CO_2 + \text{Light energy} \rightarrow C_6H_{12}O_6 + 6O_2$$

$$C_6H_{12}O_6 + 12H_2O + \text{Light energy} \rightarrow 12H_2 + 6CO_2$$

Example: Blue-green algae (Cyanobacteria) Cyanobacteria that produces hydrogen can either be nitrogen fixing (*Anabaena* sp) or non-nitrogen fixing organism (*Synechoccus* sp).

Photofermentation

It is a fermentative conversion of organic substrates into hydrogen and carbon dioxide by using sunlight as the source of energy. Under anaerobic conditions, Purple Non Sulfur bacteria (PNS) are able to use simple organic acids as electrons donors reduce proton into hydrogen gas

$$2CH_3COOH + 4H_2O + \text{light energy} \rightarrow 8\ H_2 + 4CO_2$$

Some of the PNS are *Rhodobacter sphaeroides Rhodobacter capsulates Rhodobacter palustris* and *Rhodospirillum rubrum.*

Dark fermentation

It Involves the production of hydrogen in a dark environment without the presence of sunlight, water and oxygen. Fermentative / hydrolytic microorganisms hydrolyzes complex organic polymers to monomers which are further converted to a mixture of lower molecular weight organic acids and alcohols by necessary H_2 producing acidogenic bacteria.

$$C_6H_{12}O_6 \rightarrow 2\ CH_3COOH + 2\ CO_2 + 4\ H_2$$

Some of the fermentative microbes are *Enterobacter aerogenes, Enterobacter cloacae, Clostridium butyricum* and *Clostridium acetobutylicum.*

Bio Methane Production

Biogas is an alternative and renewable energy source produced through the anaerobic (oxygen free) digestion of organic matter whereby the organic matter is converted into a combustible biogas rich in methane (CH_4) and a liquid effluent.

In general, biogas consists of 55 percent to 80 percent methane and 20 percent to 45 percent carbon dioxide (CO_2). However, depending on the source of the organic matter, small amounts of other gases such as ammonia (NH_3), hydrogen sulfide ($H_2 S$), water vapor (H_2O) and other trace gases may be present. It is the methane component of the biogas that will burn or produce energy.

The decomposition of the organic material can be divided into four different phases.

Hydrolysis stage

In the first phase, complex long-chain structures such as carbohydrates, proteins or fats are broken down into smaller molecules such as amino acids, sugars and fatty acids.

Acidogenesis stage

During the subsequent acidogenesis, the intermediate products formed in the first step are further degraded to form lower fatty acids and other carboxylic acids, such as butric and propionic acid. In addition, hydrogen, carbon dioxide and acetic acid are produced, which are the basic elements for the subsequent production of methane.

Acetogenesis stage

During the third phase of the decomposition process, referred to as acetogenesis, the lower fatty acids are broken down into acetic acid by acetogenic microorganisms.

Methanogenesis stage

Thus, the starting products are available for methanogenesis from the acetogenesis, the fourth step of the biogas process. During this phase, archaea - the oldest forms of life on earth convert acetic acid or hydrogen and carbon into methane.

Methane production is carried out by the highly specialized and obligately anaerobic methanogenic prokaryotes, all of which are archaea. Methanogens use carbon dioxide as their terminal electron acceptor and receive electrons from hydrogen gas (H_2). They produce methane and others *viz.*, methanol, formic acid, acetic acid, and methylamines.

Some of the Methanogens are *Methanobacterium, Methanococcus, Methanosarcina, Methanospirillum.*

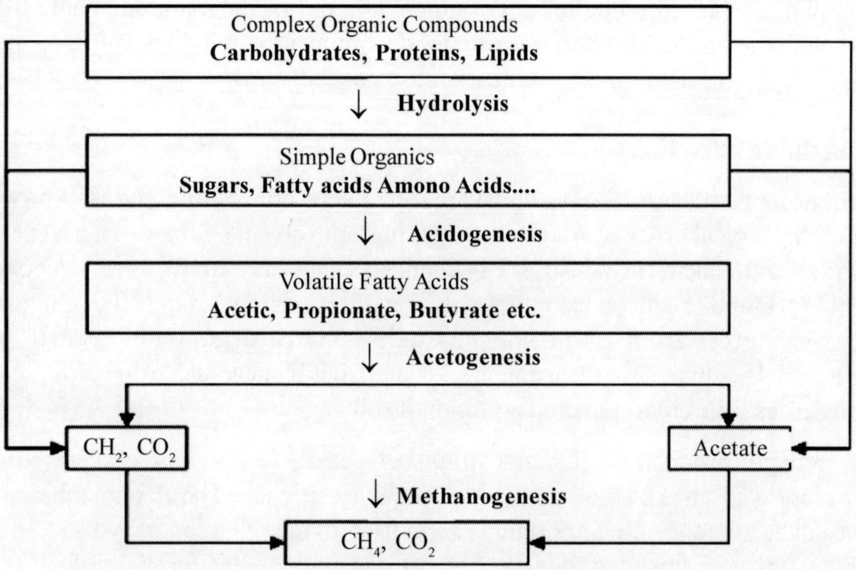

Fig. 1: Simplified overview of the AD process; adapted from Drosg, 2013

Biomethane is as a green non-fossil source of energy. Biomethane is produced from biogas derived from organic matter (often from sewage, landfill, food waste or distillery waste) which makes it a renewable source of energy. Biogas can be dried and cleaned to remove impurities and upgraded to pure biomethane. Biomethane is the fuel that is produced by refining and removing any impurities from the biogas. And unlike fossil fuels, which are considered a finite resource, the natural gas produced from these sources is a renewable resource.

Biomethane is simply biogas which has been cleaned up, or "upgraded". During upgrading the contaminants in the raw biogas stream are absorbed or scrubbed, leaving substantially more methane per unit volume of gas. Traditionally there were four main methods of upgrading: water washing, pressure swing adsorption, selexol adsorbtion, and amine gas treating. Nowadays, there is increasing use of gas separation membrane technology to perform this function.

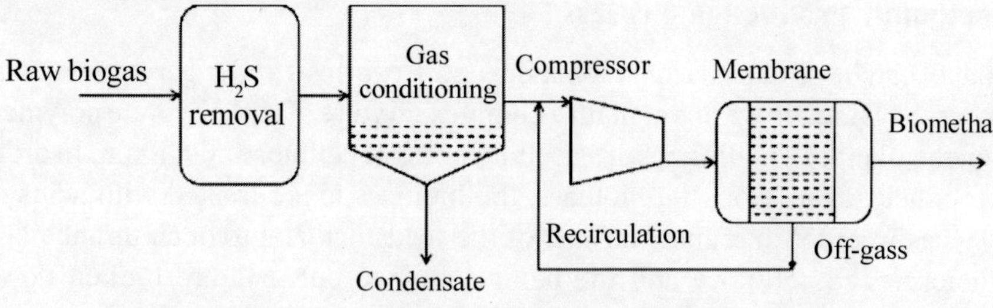

Adapted from Masood Rezaei and Dasht Arzhandi, (2017)

Biomethane has a number of environmental benefits, the main one being that the methane produced by organic matter decomposition would be released into the atmosphere if simply left to decompose naturally.

Bioethanol Production

Ethanol production from sugary materials is one of the oldest known microbiological process. Alcohol is an important solvent and raw material used in a variety of chemical industries.Bioethanol is chemically known as ethyl alcohol (C_2H_5OH) and produced from fermentation of fermentable sugars (i.e. glucose, sucrose, etc.) from plant sources using micro-organisms (yeasts or bacteria). Bioethanol is a clear colourless liquid, it is biodegradable, low in toxicity and causes little environmental pollution if spilt.

For ethanol production, selected strains of Yeast (*Saccharomy cescerevisae*) are employed since all the strains are not equally efficient. The alcohol tolerance and sugar tolerance are important criteria used in the selection of yeast strains. The yield of ethyl alcohol is about 50 per cent of the fermentable sugar concentration. Further purification of ethyl alcohol is done by fractional distillation. In some distilleries, the yeast is recovered and used as animal feed while in most, it is discarded into the effluents, a procedure that is very undesirable.

Raw materials

The widely used raw materials for ethanol production are Wheat, Maize/Corn, Sugar Beet and Sugarcane.

Microorganisms for Bioethanol Production

Yeast: *Saccharomyces cerevisiae, Pachysolen tannophilus, Candida shehatae, Pichia stipitis*

Bacteria: *Zymomonas mobilis, Bacillus stearothermophilus, Escherichia coli, Klebsiella oxytoca*

Bioethanol Production Process

Ethanol can be produced from biomass by the hydrolysis and sugar fermentation processes. Biomass wastes contain a complex mixture of carbohydrate polymers from the plant cell walls known as cellulose, hemi cellulose and lignin. In order to produce sugars from the biomass, the biomass is pre-treated with acids or enzymes in order to reduce the size of the feedstock and to open up the plant structure. The cellulose and the hemi cellulose portions are broken down (hydrolysed) by enzymes or dilute acids into sucrose sugar that is then fermented into ethanol. There are three principle methods of extracting sugars from biomass. These are concentrated acid hydrolysis, dilute acid hydrolysis and enzymatic hydrolysis.

Sugar Fermentation Process

The hydrolysis process breaks down the cellulostic part of the biomass or corn into sugar solutions that can then be fermented into ethanol. Yeast is added to the solution, which is then heated. The yeast contains an enzyme called invertase, which acts as a catalyst and helps to convert the sucrose sugars into glucose and fructose. The chemical reaction is shown below:

			Invertase			
$C_{12}H_{22}O_{11}$	+	H_2O	→	$C_6H_{12}O_6$	+	$C_6H_{12}O_6$
Sucrose		**Water**	**Catalyst**	**Fructose**		**Glucose**

The fructose and glucose sugars then react with another enzyme called zymase, which is also contained in the yeast to produce ethanol and carbon dioxide. The chemical reaction is shown below:

		Zymase		
$C_6H_{12}O_6$	→	$2C_2H_5OH$	+	$2CO_2$
Fructose/Glucose	**Catalyst**	**Ethanol**		

The fermentation process takes around three days to complete and is carried out at a temperature of between 25°C and 30°C.

Fractional Distillation Process

The ethanol, which is produced from the fermentation process, still contains a significant quantity of water, which must be removed. This is achieved by using the fractional distillation process. The distillation process works by boiling the water and ethanol mixture. Since ethanol has a lower boiling point (78.3C) compared to that of water (100C), the ethanol turns into the vapour state before the water and can be condensed and separated.

Fuel properties of bioethanol

Ethanol can be blended with gasoline to create E85, a blend of 85% and 15% gasoline. Due to the corrosive effects of E85, because of its high alcohol content, traditional vehicles cannot use it. It needs flex fuel vehicles. Flex fuel vehicles (FFVs) have engines modified to accept higher concentrations of ethanol. Such flexible-fuel engines are designed to run on any mixture of gasoline or ethanol with up to 85% ethanol by volume. Bioethanol has much lower energy content than gasoline. About two-third of the energy content of gasoline on a volume base. Octane number of ethanol is higher than petrol, since ethanol has higher antiknock characteristics. It increases the fuel efficiency of the engines. Also ethanol has the oxygen content of 34.8 %, which also leads to a higher efficiency, which results in a cleaner combustion process at relatively low.

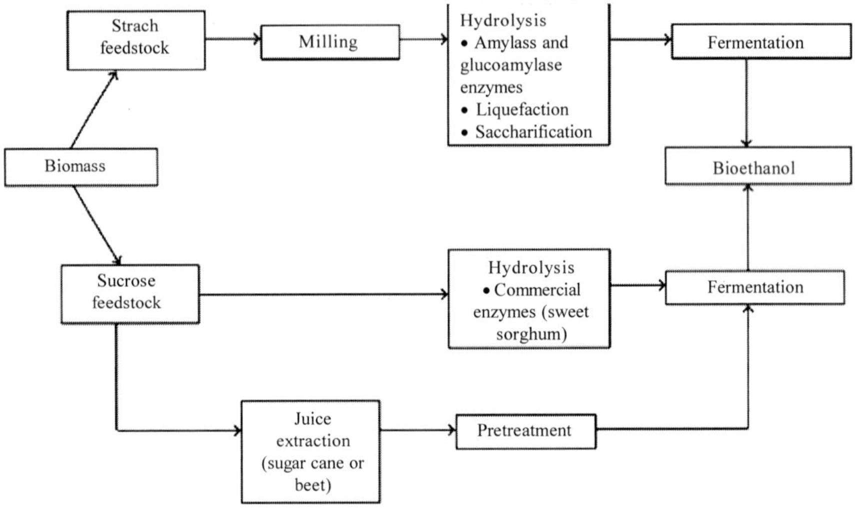

Blending with a small proportion of a volatile fuel such as gasoline is more cost effective. Various mixture of bioethanol with gasoline or diesel fuels are used.

E5G to E26G - (5-26% ethanol, 95-74% gasoline)

E85G- (85% ethanol, 15% gasoline)

E15D - (15% ethanol, 85% diesel)

E95D- (95% ethanol, 5% water, with ignition improver)

Benefits of bioethanol

- It is a renewable fuel
- Less dependence on crude oil
- Increase octane number (a standard measure of the performance of motor fuel)
- Reduces air pollution, cleaner environment due to cleaner combustion emissions
- Lower net carbon dioxide
- Bioethanol also can serve a wide range of uses in the pharmaceuticals, cosmetics, beverages and medical sectors as well as for industrial uses.

Biodiesel production

Biodiesel is an alternative fuel to conventional or fossil diesel. Biodiesel is defined as non-petroleum-based diesel fuel consisting of alkyl esters (methyl, ethyl, and propyl) of long chain fatty acids. Biodiesel, a mixture of fatty acid methyl esters (FAMEs) derived from animal fats or vegetable oils. It is produced by transesterification reaction of oil from plants (such as soybean and jatropha), animal fats, and from microalgae or fungi.

Biodiesel production using microbial lipids, which is named as single cell oils (SCO), has attracted great attention in the whole world. Single cell oils (SCOs), are produced by oleaginous microorganisms including algae, fungi and yeast, and are considered to be promising alternatives to vegetable oils as their oil properties are similar.

Oleaginous microorganisms

Lipids (oils), proteins and carbohydrates are the three macromolecules occurring naturally in microbial systems. Lipids are an important structural component of the microbial cell membrane. Certain microorganisms (algae, yeast, fungi and bacteria) can accumulate more than 20% of lipids inside their cells as triacylglycerol (TAG). Those microorganisms are called as oleaginous microorganisms. The oils thus produced are popularly called SCOs (single cell oils).

Algae

Microalgae are good candidates for biodiesel production. They have faster growth and higher biomass production. They live in harsh conditions due to their unicellular or simple multicellular structure. Their average lipid content is 1-70% with a maximum of 90%. Some of the common algae are *Chlorella, Crypthecodinium, Cylindrotheca, Dunaliella, Isochrysis, Nannochloris, Nannochloropsis, Neochloris, Nitzschia, Phaeodactylum, Porphyridium, Schizochytrium, Tetraselmis*.

Bacteria

Bacteria can accumulate oil of about 20-40% of dry biomass. They have superiority in the production of biodiesel due to their highest growth rate.

Acinetobacter calcoaceticus - 38%

Arthrobacter sp. - 40%

Fungi & Yeast

They are considered as favourable oleaginous microorganisms. *Rhodosporidium* sp., *Rhodotorula* sp. and *Lipomyces* sp. can accumulate intracellular lipids as high as 70% of their biomass dry weight. *Cryptococcus curvatus*- Oleaginous yeast accumulate storage lipid >60% on a dry weight basis.

Biodiesel production from microalgae

The production of biodiesel from microalgae has multiple advantages and has been termed the third-generation biofuels. Unlike other oil crops, microalgae grow extremely rapidly and many are exceedingly rich in oil. Microalgae commonly double their biomass within 24 h, and biomass doubling times during exponential growth are commonly as short as 3.5 h.

Culturing of microalgae

The large scale production of microalgae is generally performed with solar energy (photoautotrophic metabolism) in open ponds (raceways), closed systems (photobioreactors) or fermentors.

Harvesting of microalgae

Microalgae biomass could be harvested by, centrifugation, flocculation, gravity sedimentation, filtration methods.

Extraction of lipids

The lipids are extracted using chemical solvents, chloroform-methanol blends have been extensively used with high extraction yields up to 83%. Some physiochemical as wellkas biochemical techniques are used for extraction of lipids.

Transesterification

It is the process of converting oils to biodiesel. In this process lipids can react with alcohol and produce biodiesel.

Purification of biodiesel

Biodiesel and by-products must be separated for increasing the biodiesel production. The main separation processes used hot water (50°C), organic solvents such as hexane and water organic solvent for a liquid-liquid separation.

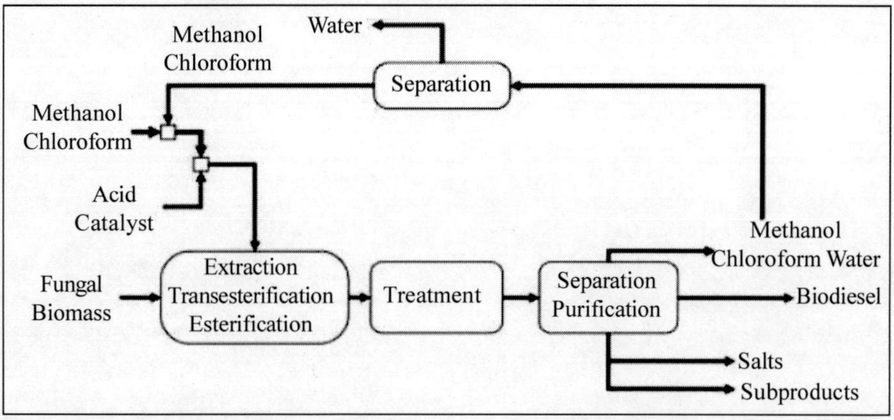

Advantages of Biodiesel

Biodiesel has many environmentally beneficial properties. The main benefit of biodiesel is that it can be described as 'carbon neutral'. This means that the fuel produces no net output of carbon in the form of carbon dioxide (CO_2). Biodiesel fuel is a renewable energy source, unlike petroleum-based diesel. It is less polluting than petroleum diesel. The lack of sulfur in 100% biodiesel extends the life of catalytic converters. Biodiesel can be blended with other energy resources and oil.